Hidden History, Rain Engineering and UFO Reality

The Best of Trevor James Constable

THE BOOK TREE
San Diego, California

ISBN 978-1-58509-148-5

Cover Images

Cover layout
Mike Sparrow

Retyping and editing,
various chapters & historical pieces
Janice Boles

Published by
The Book Tree
P O Box 16476
San Diego, CA 92176
www.thebooktree.com

Contents

Contents

INTRODUCTION

What follows is a collection from my life's work, which centers mostly on the development of successful rain engineering devices of various designs. These weather altering devices have solved drought problems in places around the world. Variations of them, called cloudbusters, have attracted interesting – and otherwise hidden – life forms in the sky which were documented by myself and others with infrared film. The main purpose of my work, however, has been to help mankind when in need of its most vital resource – water.

Our learning and development process with rain creation was lengthy, expensive and often difficult, but it was essential preparation for going airborne with a technology that totally confounds our brightest minds by making a quantum jump into the ethers. No such exciting leap has previously occurred in the world's study of meteorology or in any aspect of industry related to this field.

Those who come after us will now find it much easier to advance the art. The situation is akin to aviation itself. In which the youth of today's world can leap right over the pioneering of the Wright brothers and the development of aero engines and aeronautical technology. All the crashes and the losses of valiant pioneer pilots like Kingsford Smith, Wiley Post and Amelia Earhart, who flew into the void never to be seen again, make up the historical anatomy of air power. Mighty passenger jets leave us gaping, as they cross our oceans laden with hundreds of humans. This is only the beginning.

Be prepared for the etheric age that is now upon us. We are entering the age of the ethers. This development cannot be stopped, only stalled temporarily, by ignorance or malice. Our tiny pioneering group has cleanly and clearly opened the way to a new and unpolluting source of power. This power cannot be put into a wire or a tank and sold. This is the power of the ethers, Nothing can stop humanity's march to mastery of this natural force any more than the numerous barriers we surmounted stopped us. The biggest single barrier lies within man. Free primary power awaits humanity's enlightened touch. We have had our fill of dealing with evasive and frightened politicians, greedy capitalists and others who sought to take commercial advantage of our findings while cutting us out. They did not understand that etheric power is for all mankind, to lift the downtrodden and inspire the brave and power the kind among us to be even more generous to our fellow humans.

The late Dr. Wilhelm Reich found his way to the primary force behind everything living. He called that force in 1939-40, orgone energy. Dr, Reich's pathway to this energy was via the first scientific study ever made of the human orgasm, the primal convulsion that put us all here. Until Dr. Reich turned his genius on sexuality and our origins, the human race knew far more about explosives, and how to blow human beings into bits, than it did about humanity's roots in a heretofore unknown form of energy – the life energy.

You may, indeed, turn away from the etheric age, and thus "apply the brakes" to your own further unfoldment. That will not prevent the ether from pursuing its destined course. You cannot prevent your ultimate involvement with the ether. Better by far to begin your emancipation from all the ills and diseases that originate in trying to lock out the ether from your functioning. You can begin your emancipation by following my steps in pursuing the etheric control of this planet's weather. That rugged pathway has been made easy for you by this book, which describes how I arrived at my ultimate fulfillment – airborne etheric rain engineering. This radical technical development, made without chemicals or electric power in any form, will excite the hostility and malevolence of academia for some time to come. Ultimately, the cosmic point of view always wins out. Why not join us in this non-political revolution?

Look at the caliber of those who put their shoulders to ours through the years. George Wuu, the Singapore millionaire, the late Frank Der Yuen, internationally esteemed aeronautical consultant; businessman Howard B. Morrow; Air Force General Curtis LeMay, and on into the dozens of good citizens who helped us in our labors. Their contributions are outlined on our website, rainengineering.com. We would not have this book today without them.

Despite the triumph of airborne etheric rain engineering, our tasks are not yet complete, mainly because so many people in high office remain paralyzed by the portent of the ether's entry into a world dominated by profit. Thus, the way ahead appears fraught with difficulties, just like the past. This means that we must continue to work not just for the improvement of the technology, but also for the improvement of public attitudes that will welcome the ether and a New Civilization that awaits mankind. All these things can be done, if the basis for it is understood. That is the purpose of this book, which details my life's work with this profound and powerful energy. This book will allow you to understand this energy, which waits at our fingertips – including the world of etheric rain engineering and the strange etheric creatures that we've stumbled across in our quest to both harness it and understand.

I have also included in this collection two articles about little known military men who changed history – Brigadier Ralph Bagnold and Major-General Percy Hobart. Although not quite in line with the rest of the book, subject-wise, I found these men to be of such historical importance, that they had to be included. Unsung heroes of this magnitude should have whatever available credit extended their way.

I hope you find this work, which consists of the very best I have done in my life, to be fascinating and inspiring. It is my continued hope that what I've accomplished can be a good starting point for helpful human endeavors.

Trevor James Constable

Chapter 1

ETHER, ORGONE ENERGY AND CHI
Reflections of a Weather Engineer

Terminological confusion encumbers and impedes discussions of weather engineering, and frequently confounds scientists. Why is this so? The main reason is that the late Wilhelm Reich, MD, added *orgone energy* to Western thinking in electricity and kindred fields in 1939-40. He characterized it as a specific biological energy, and official science resisted this addition to the energy armamentarium. Official science still resists orgone energy, seventy-five years after its discovery.

No effort was made by official Western science to investigate Dr. Reich's lengthy clinical and practical career, which included his service as First Clinical Assistant to Dr. Sigmund Freud. From those days to these, all of official medicine appears stupefied when called on to explain the difference between a living human being and a corpse. Dr. Reich devoted his life to the detectable energic presence in a living human that is no longer present in a corpse. Official science took the silent way out, tacitly agreeing with official psychiatry that Dr. Reich was crazy. For his stellar discovery, Reich's reward was to be misrepresented and misunderstood the rest of his life, culminating in his death in an American Federal prison.

Convinced through long experimental experience that he had identified the energy of *life itself*, Dr. Reich was compelled by a storm of opposition to found a new and independent scientific discipline, which he called *orgonomy*. He took this step to protect and develop further what he had discovered as a universal presence in the living organism, in the soil, in the atmosphere, and in every living thing from the microbe to the mastodon. He chose the name *orgonomy* to relate the energy to things organic, to the organism, and to the orgasm, in which primal convulsion we all have our physical roots.

Through his extensive clinical experience with psychoanalysis, Dr. Reich had learned many nasty things about his fellow man. Especially was he aware of what normal-seeming men hid behind their many masks. His work had to be protected from such men, because Dr. Reich knew that they could only react to his life-giving breakthrough by seeking to kill that discovery.

Wilhelm Reich

Dr. Reich published his experimental methods in special *Orgone Energy Bulletins* for other scientists to study and replicate. He wrote several books, including the monumental two-volume study, *The Discovery of the Orgone*. He left behind after his death, approximately 100,000 pages of manuscript, enough for ten more books. Those who have dreams of becoming writers may marvel at the magnitude of this legacy. There was, and there is therefore, nothing whatsoever secret about orgonomy. Far from it.

Thousands of years before America was even thought of, ancient China's physicians had recognized what Dr. Wilhelm Reich, centuries later, would call orgone energy – the life force. The ancient Chinese doctors called this force *chi*. They still call it chi. The existence of flows of chi along meridians of the human etheric body, to sustain and maintain the various organs, glands and tissues, are an integral part of a Chinese physician's education. This approach is highly developed and well understood in China. Chi remains outside Western medicine, which has been overwhelmed and largely subverted by profit-driven, drug-based medicine.

Traditional Chinese medicine, or TCM, is the art of correctly directing the CHI to eliminate blockages to the chi's proper flow. Chinese herbs, rightly understood and applied, bring cosmic forces into the healing process. Many years are required to understand and apply these advanced herbal therapies. Three thousand years ago, China already had a complete pharmacopoeia, based upon normalizing chi flow. The Chinese doctor thus freely uses in his healing art the same forces of chi that landed the distinguished Dr. Reich in a Federal prison. He passed away a week before his scheduled release in 1957.

By 1950, Reich had wearied of the seemingly endless task of treating, psychoanalytically, people with disordered psyches. Thousands of persons needed help. Civilized life ruined healthful functioning far faster than the stricken could be helped via psychiatry. Even though Dr. Reich added manipulative techniques to his treatments, which aimed at releasing orgone energy bound in the musculature, the process was accelerated somewhat, but remained a slow one. Dr. Reich began extending the bio-energetic reaches of his orgone discovery.

Dr. Reich's powers of observation were truly phenomenal, according to the late biologist, Robert A. McCullough. He worked with Dr. Reich for two years. McCullough himself was an exceptional observer and also a sensitive. He was associated closely with this writer in weather engineering for over twenty years. McCullough told this writer of the numerous times that Dr. Reich had detected with his ultra-sharp vision what he, McCullough, had completely overlooked.

Dr. Reich's observational powers were soon drawn to the surface of nearby Lake Mooslookmeguntic in Maine. Distinct "heat waves" appeared above the lake's surface. He returned repeatedly to these "heat waves," which are more properly known as "atmospheric boil." Reich noted that there was a distinct directional component to this atmospheric boil, and it ran mainly from west to east. He asked himself, what could be the cause of this undeniable directional effect?

Regular observation ruled out the wind as the motional cause. Sometimes, Dr. Reich would note that the atmospheric boil would reverse its direction of flow. He further noted that the reversal happened when there was a large low-pressure system lying west of Maine. Could all this be due to a vast regional flow of orgone, or CHI? Dr. Reich kept studying these motions, without reaching any firm conclusion, until a small accident one day made a valid physical connection possible. This connection portended vast potential, including revolutionary revisions of physics.

Construction was in progress in the buildings of the Reich facility in Maine. In a hollow in the ground nearby, a jumble of pipes, lumber and other construction items had been roughly stored. There were several inches of rainwater in this hollow, on whose edge Dr. Reich stood while making his observations of the lake. On this occasion, intent on his observations, he missed his footing and stepped down on to the pipes in the hollow.

This caused one pipe to rise up as its opposite end was pressed down into the rainwater in the hollow. Dr. Reich was astonished to see the atmospheric boil above the lake's surface – far out on the lake – jump upward as though in response. He again pressed the crucial pipe end down into the water in the hollow. Again the atmospheric boil far out on the lake jumped in response. Repetitions confirmed the connection.

This was action at a distance. Reich saw its significance, his genius aroused by the obvious possibility of controlling the orgone flow, and thereby controlling the weather. The earth's orgone energy (chi), in a massive west-to east regional flow in the Maine latitude, could actually be reached, and influenced, by a pipe grounded at one end into water. A single pipe could influence that flow. He had done that already. Dr. Reich's enthusiasm for bio-energetic research had presented him with an immense challenge: exploring weather control by tapping into the regional flow of orgone energy (chi).

Perhaps that flow could be built up, diminished or diverted in some way. First he would need to use his powers of observation to establish something of the flow laws of orgone (chi). He would need to design a device consisting not just of a single pipe, but of an array of pipes. He could ground this device into water and use it to access the observable skies. Thus he could begin to probe the physics of this regional energy flow. He could possibly find out, as a practical matter, just how much activity could be controlled – and perhaps engineered! Thus was born the device that entered history, and began an all-new science – the cloudbuster.

What Dr. Reich conceived, in an astounding prevision of genius, involved no arcane circuitry. A ten-dollar transistor radio far exceeds the cloudbuster in complexity. Five telescoping metal tubes were mounted side-by-side, in a two-over-three pattern, on a rotatable platform. One end of each tube was attached to a flexible BX cable that could be lowered into water, preferably flowing water. The tubes could be elevated from the horizon to the zenith, or set at any intermediate point. He now could access the entire skyscape with his invention.

Rapidly, Dr. Reich found out that if this assembly were aimed at a discrete cloud, that cloud would in due course disappear. That is, the device would "bust" that cloud. Hence the name, "cloudbuster." Dr. Reich similarly discovered that if the cloudbuster were aimed near a cloud, rather than directly at it, the cloud would in due course expand and fill in the space between itself and the point of aim. Dr. Reich carefully made time-lapse films of both the busting and augmentation functions.

In overcast conditions, leaving the cloudbuster aimed directly upward, *holes* in the overcast would eventually appear. The pattern of the holes replicated the two-over-three mounting of the tubes on the cloudbuster. There was no chemistry or electric circuitry involved in performing these extraordinary functions. Dr. Reich had begat a wondrous instrument.

This writer has designed, built and used more than twenty cloudbuster devices of various kinds. Ranging from straight Reich-type units down to grossly simpler, ungrounded geometric units that can be taken inside light aircraft or helicopter cabins, or attached to their wing struts or landing skid assembly, they all worked. Cloudbuster functions are the future calling to man, and such units are not made out of dream stuff.

The writer has countless hours of time-lapse video verifying that the technology founded by the genius of Dr. Reich is alive and healthy and completely proven. The cloudbuster was a monumental breakthrough to what could now fairly be called *chi engineering*, should the Chinese elect ever to harness this primary force for purposes other than medicine.

THERE CAN BE NO DOUBT IN THE MIND OF ANY EXPERIENCED PRACTICAL WORKER THAT THE CHI OF ANCIENT CHINA, AND THE ORGONE ENERGY OF DR. WILHELM REICH ARE IDENTICAL.

Since the chi is universal, the chi flow in Wilhelm Reich's Maine is also the chi flow of China, and of Australia, Africa and Russia. Chi and its energy flows in all countries. The chi or orgone flows without asking any doctors' or scientists' or politicians' permission. No learned society controls or authorizes chi or orgone.

Updated October 9, 2008

Chapter 2

THE COSMIC PULSE OF TREVOR JAMES CONSTABLE

Interview by Joan d'Arc

(This exclusive interview took place by telephone in 1994)

Plasmoids That Live in the Sky might be the title of a campy fifties sci-fi movie, but in this case it's the subject of the life work of Trevor James Constable, and the result of years of scientific research gathered and corroborated worldwide and presented in his book *The Cosmic Pulse of Life*. When he is not documenting the existence of macro-bacteria on infrared film, Trevor is doing some serious "cloudbusting" experiments, explained in his video *Etheric Weather Engineering on the High Seas* and in a forthcoming book. Here, Trevor talks about fringe science, the invisible realm and the flummoxing of earthling power elites by evil entities having no business doing business in this dimension.

Joan d'Arc: Mr. Constable, I read that your book, *The Cosmic Pulse of Life*, first published by Merlin Press in 1976, which "shocked the publishing industry" because of its criticism of official science – a "science" which has been unable to make a significant finding in the area of UFO phenomena in over forty years.

Trevor James Constable: Well, first of all, the shock which is alleged to have occurred in the publishing world is, I think, exaggerated. I think it had a serious impact on the UFO world, because people suddenly found that a large mass of evidence had been gathered, and it was original evidence. It hadn't been put through any kind of government or scientific strainer where the life could be knocked out of it. This indicated that the UFO thing was something immeasurably more complex than just a bunch of entities climbing into spaceships over on Venus or Mars or somewhere else in the universe and traveling over here to take part in the life of the Earth in some incomprehensible fashion. It showed that there was a biological dimension to the whole thing, and this in turn is the key to the general sequestration of the whole approach of *Cosmic Pulse of Life*. It is true that, since the days of Kenneth Arnold, when

UFO's first entered the public domain, there has been no single significant finding in the area of UFO phenomena.

d'Arc: Yet, answers seem to be closer when working with the theories of Wilhelm Reich and others considered to be in the realm of "fringe" science.

Constable: That is a fact. It is true that answers are closer when you work with the theories of Wilhelm Reich. Because you must remember that Reich's work is rooted in practical matters. And his discovery of this biological energy, a specific biological energy that he called *orgone*, set the entire question of vital force on a completely new basis. Up until then, it had been sort of a philosophical will-ó the wisp. Reich changed all that, and by using modern instrumentation and modern methods he presented evidence of the existence of this biological power, the thing that makes every heart beat and which nobody, but nobody, in the entire reaches of official science has yet seen fit to address.

d'Arc: Why do the High Priests of Science need to protect the materialist worldview at all costs?

Constable: [Imagine being] raised from infancy with chains of interlocking and interlacing views laid upon you and made a part of your method of functioning and viewing creation, and all this excludes the idea of a biological energy; there isn't any pulsation, it's just chemicals. Then all of a sudden someone comes along and says, "Hey, wait a minute, there is a bio-force!" And it's physical; it's accessible. You can demonstrate it! Well, what happens is everything that you know, everything that you have been taught, everything that makes up the cement of your whole worldview is subject to a very serious disillusionary attack. That is the reason why they clamp down in official scientific circles at the very suggestion of specific biological energy, orgone, or as we refer to it in my phase of weather engineering, chemical ether.

d'Arc: Explain your theory that some UFOs are "plasmoidal living organisms" which are native, albeit invisible, to our atmosphere and how you came to realize that these biological entities could be photographed using infrared film.

Constable: As explained in my book, when Dr. Woods and myself started this work back in 1957, we went after spaceships like everybody else. We had no consciousness whatever of there being anything else. And when we used infrared film for the purpose of pursuing objects that were repeatedly seen and reported worldwide as having come from and often returned to an invisible state, we recorded not so much spaceships as these strange plasmoidal living organisms. And there were too many of them to be ignored, or blown away as aberrations or the results of faulty processing: the normal excuses that are made for these things.

And, we were not at all happy at the time that we obtained the first evidence of the plasmoids – not at all, because we wanted spaceships. We were in that same hard, mechanical idea about these things, that someone had built these things over in Venus or Mars or elsewhere, and they jumped in them and flew over here, and they were cheek by jowl with us on Planet Earth. At least that's what we wanted to be able to prove.

In actual fact, using infrared film, which extends the range of the human eye far beyond what the eye itself will register, we got something else. We got these living organisms, and we got lots of them. And since that time lots of other people all over the world have got them. I just got a letter from a gentleman who photographed shoals of these things over the Western Australia desert. So, they are there and they have always been there. They are not from anywhere else; they are what we might call the upper dimension of our physical world. They have been referred to as macro bacteria: the counterpart at the upper border of physical nature of the microbes that infest the sub-sensible realms of physical nature of which we were earlier not aware. We are aware of them now, because centuries ago the microscope was discovered and an optical limitation was removed. Infrared film does the same thing at the upper border of physical nature, but the things that are found, instead of being extremely small, they are very vast, sometimes forty or fifty feet across.

So everything fits from a polarity point of view, and finding out that these things could be recorded with infrared film was simply accidental. It was not something that was planned, programmed or intended. But we didn't shrink from what we found, we pursued it through the years and we never found any reason to revise it. The work that was done subsequently in Italy twenty or thirty years later served only to verify it, as has the work done by other people in Romania, Australia and elsewhere. All through the years, all of this has been mercilessly crushed by the people in the UFO field who feel that they are somehow serving the truth by killing something that's alive.

d'Arc: How does orgone energy (" the ether") relate to your exobiology theory and your work in Reichian cloudbusting?

Constable: I would need a three or four hundred page book to describe my work in Reichian cloudbusting. In fact, a fairly comprehensive volume on this is being prepared now and is going to be published by Borderland Sciences Research Foundation sometime this year. So it is futile to attempt an interview at this time to describe my work. I haven't done Reichian cloudbusting in a while; I do things in a completely different way, I use geometric forms. I do not use the old time Reichian cloudbuster anymore, although I made good use of it for many years.

I can only say it is impossible to assemble any kind of valid exobiology theory if you do not first understand why it is that everything living on this Earth, from the microbe to the mastodon, pulsates! That's the thing that you have to elucidate first.

d'Arc: If Reich had lived beyond the mindset of the fifties, do you think he would have changed his view of UFOs as "machines from another planet?"

Constable: Sure, I think that he would have readily expanded his initial conception of UFOs as machines from another planet. However, I want to point out something significant here, and I never tire of re-emphasizing this, because it never seems to sink in, that it's not a question of having spaceships or plasmoidal organisms: you have both! There are at least two dimensions to the UFO phenomenon. One is the spaceships idea and the other is the living organisms. And because in the mode of their manifestation these two aspects of the UFO phenomenon are mutually confused, nobody seems to have been able to straighten it out, except me.

People believe that you've got ships from other planets and you can't have anything else. If you're going to have biological critters in the upper atmosphere, well then you can't have the other. You've got to have both. And what this does is expand your consciousness. And that is one of the good things about this whole phenomenon in all its diversity. You've got to get your consciousness expanded and operating along new lines. The old ways are dead and gone, like this civilization of ours. I certainly believe that Reich, had he lived, would have expanded his understanding of these things, because he had the mentality and the observational power to do so. But, you can't expect very much of a man at a critical time of his life when he was being harassed and hounded by the medical profession, the psychiatric profession, the FDA, and the courts. You can't expect much from a man who is under that kind of pressure.

d'Arc: What was your theory of space and flying saucer propulsion in 1958, when you wrote your classic book *They Live in the Sky*, and how have your theories changed since then? Were you influenced by Wilhelm Reich's work (described in his book *Contact With Space*) and were you pursuing a similar path in your research?

Constable: My theories in 1958 I cannot recall for you in detail or compare them to the views that I hold now. But basically there wasn't a great deal of difference, it's just that in the meantime more and more evidence has accumulated. The basic idea that the source of all these things is the invisible has continued to gain momentum and continued to gain status because of the derelictions and inadequacies of the formal theories. I was influenced by Wilhelm Reich, but not until after I had made my original contribution with *They Live in the Sky*. I did not know Reich existed or I might have written that book differently. That influence came along later and is part of subsequent development worked out in *The Cosmic Pulse of Life*, which correlates with work from Romania and Italy.

d'Arc: Does the appearance of UFOs in the skies have anything to do with the "destruction of the industrial, financial, moral and spiritual fabric of the USA?" Is the American decline an engineering job on the part of an otherworld intelligence, which somehow rules the planet Earth through its top power elite?

Constable: Everything that has happened and everything that we are seeing now that shows this ongoing downtrend in American Life, the industrial, financial, moral, spiritual and political fabric of the USA disintegrating, this is all being brought about by people who are in the body with us. There is no question about that. This is a function of conspiratorial activity. It is the inspiration of it that comes from other planes; the guidance of it. And very often these people have no conception whatever that they are under psychic control and are being directed from other planes to do these things by entities who cannot manifest here themselves. That, in my opinion, is the reality of any infiltration of the top power elite, which is, by the way, not something visible to the ordinary citizen. It is hidden also, as most people know, down at the level of the Presidency. Somebody like George Bush (Sr.), for example, would be probably at about the fourth layer down in the power structure. The entire world is run from the shadows and it is well to keep that in mind.

d'Arc: In your book you state that scores of people believe there is some undetermined energy principle behind the UFO phenomena, yet the billion dollar NASA budget does not contain a dime for UFO investigation. Why does the government bury its head and continue to pretend there is nothing out of the ordinary happening in the Earth skies?

Constable: Of course it is true that there is an undetermined energy principal behind the UFO phenomena, and NASA has no hesitation in squashing billions, which it doesn't have to earn, and spending nothing to investigate UFOs or even have a small project on them. Nothing. The reason for it is very complex and relates to the energy question. We know from Reich's work that this energy will run motors, and if you can run motors out of the atmosphere without having to hook into the power grid, that is a disaster for the world's power structure. The control of energy is the control of people. It is the control of economies. So the breaking of the fuel monopoly is reason enough for governments who are run from the shadows and steeped in corruption to do nothing to elucidate the principles of energy that are behind UFOs.

d'Arc: Can you discuss your statement that "unseen extra-physical entities have expanded their manipulations and destructive capabilities on the physical plane via humans they can control." Do you see this as an occult influence in top levels of world secret government?

Constable: Extra-physical entities with an interest in the earth plane can only manifest here through correspondences. They do not come down and land on the White House lawn, because they do not belong to the Earth plane. So what they have to do is find correspondences within those in authority and control what they can access and, by those means, infuse their malevolent ideas and purposes into the civilization. On the simplest and broadest basis you may ask yourself why it is that the US, despite having had the most successful and largest expansion of production in the history of the world in World War II, grew by 1950 to become a phenomenally prosperous country, and since that time has been run down mercilessly to a shadow of its former self, incapable of carrying out even many of its own defense processes, can't forge its own aircraft carrier anchor chains, stuff like that. Straight down from a high point. Despite the prodigious proliferation of computers and other things that are supposed to improve productivity and increase the standard of living, our standard of living goes down. The chief agency and the chief link between the ruling authorities, the real rulers here, and those in the extra-physical realm who are utilizing them, is the dishonesty of the men on the physical plane. That's what opens them to access from the unseen.

d'Arc: In your book you talk about Hitler and the occult influence in machinations of World War II. Most people do not believe in "magic," but is it more simply understood as a series of deals made with other-dimensional entities who shouldn't be trusted?

Constable: I referred to Hitler merely in passing and pointed out that he was essentially the vehicle of Lucifer and it was a Luciferic activity that he headed in Germany. Anyone who does not understand the difference between Lucifer, Ahriman (the evil deity in the Zoroastrian religion) and Christ will never know anything about this world. We in the USA are fully in the clutches not of Lucifer but of his opposite, Ahriman, and the Ahrimanic powers in general. This is what you are dealing with in the so-called phenomenon of the Greys, all of which is taking place in the darkness. It's a significant point; it's a signature. Nothing is out on the table, nothing is in the light, everything is in the dark.

d'Arc: So the trouble we find ourselves in is a result of bad deals made by Earth leaders?

Constable: Oh, true to form, I think that these arrangements were originally entered into with these otherworld entities for the purpose of attempting to get from them technology that could be used to make profits, technology that could be exploited and, instead, they found themselves inextricably enmeshed with entities they could neither understand nor control. And these were beings that had no trouble flummoxing them five ways from Friday. The very sound principle from basic spiritual science is to have no commerce whatever with entities of this type, to have no commerce whatever with entities who do not belong here and above all to be honest, since honesty protects you against all this. Do everything that you want to do, or that you want to see done, in the broad light of day and right out on the tabletop. That is a principle that I think our government relinquished and buried a long time ago. Alas for America!

 The Cosmic Pulse of Life: The Revolutionary Biological Power Behind UFOs by Trevor James Constable was republished by The Book Tree in 2008 (www.thebooktree.com) (800) 700-TREE.

Chapter 3

DIMENSIONS UNMEASURABLE

Excerpt from *The Cosmic Pulse of Life*

A man who has bought a theory will fight a vigorous rearguard action against the facts. — Joseph Alsop

Mechanistically-minded humans have accepted uncritically the theory of interplanetary spaceships as the fundamental explanation of UFOs. This theory has dominated the subject from the modern advent of the phenomena down to the present day. Involving only a linear projection of extant earthly technology, and doing no violence to the mechanistic cosmo-conception, this simplistic theory has paralyzed the thought processes of several generations of human beings interested in UFOs.

Acceptable as one theory relevant to certain types of UFOs, the ships-from-other-planets approach was elevated irrationally to the status of a foregone conclusion for explaining every UFO sighting. The bankruptcy of official science in the empirical phase of the UFO field is due to its having sought to prove this foregone conclusion. The time has come to establish a more rational perspective, from which we will allow the phenomena to tell us about themselves in their own way. Manifestation is a language of its own, and one we must learn. Compulsively demanding that phenomena respond to mechanistic criteria has been barren of results.

The ships-from-other-planets concept was the bedrock upon which an "establishment" in ufology was erected. Immature notions of cosmic workings characterize this approach, together with overconfidence in the current crop of mechanistic scientists. Habile and pitilessly efficient at perfecting engines of destruction, these men have drawn blanks on UFOs. The charm of the ships-from-other-planets notion lay for years in the expectation that UFOs could be understood with existing scientific knowledge, or with linear extensions of such knowledge that were deemed imminent. The approach least likely to disturb the neurotic weltanschauung, ships-from-other-planets therefore became automatically the most popular, despite its irreconcilability with a large corpus of observations.

Subscribers to this theory as the primary explanation for UFOs exhibit a marked blindness in connection with UFO propulsion. Ships-from-otherplanets as they have impinged on earth life, command a power source impenetrable to official science. Energy in some arcane form is being used for propulsion in a way that earthmen do not yet understand. Since extensions of existing technology in no way approach UFO propulsion capabilities as observed, native common sense suggests, with much evidential support, that progress may be made by taking a wholly new approach.

Ships-from-other-planets devotees usually recoil from this idea. Common sense suggests a possible beginning in offbeat, borderland areas of investigation and thought, where human beings of novel bent have always labored outside and usually beyond official science. The mavericks of this borderland include many qualified scientists who explore the field avocationally. These scientists tackle phenomena that do not square with mechanistic concepts and methods, or which seemingly spill over into the methodologically forbidden realm of faith. Such individuals are increasing in number, and they are true to the ideals of science.

Since UFOs stand outside mechanistic concepts, and have also evoked a powerful mystico-religious response among humans, this borderland area of original, untrammeled work and thought might be expected to yield valuable indications to any UFO project mounted by official science. No such approach has been made. On the contrary, scientists who had studied UFOs avocationally – sometimes investing thousands of hours of their leisure in this way – were ruled out of the Colorado University project for fear that their objectivity would be adversely influenced. Organized ufology has also been unable to extend itself even to the fringe of the borderland.

Evidence that punches holes in mechanistic conceptions – such as the multitudinous examples of materialization of UFOs – is papered over or shunted aside in favor of evidence considered "harder," that is, more accessible to mechanistic method. When the UFO subject began to take on an inevitable mystical and occult aspect, establishment-type UFO organizations responded by elevating ministers and rabbis to their boards and committees. They have proved themselves as helpless as the official scientists. The main idea in all these machinations was to maintain comfort and avoid tackling the invisible.

Contact stories were dismissed as unworthy of scientific consideration, and every establishment-type UFO organization has a list of such "cranks." The irrationality of these organizations seems incredible, since they obviously conclude that UFO intelligences should communicate only with two groups of earthmen:

1. Scientists already beaten by the phenomena, which had torn the fabric of Mechanism to rags.

2. Political leaders who were using the full machinery of government to suppress UFO evidence and discourage its discussion.

This irrational bias toward an orthodox, thoroughly safe approach to the subject has barred the way to the comprehensive theory that the facts – be they welcome or unwelcome demand out of their own stuff and substance. A comprehensive theory of UFOs cannot evade what has been observed, experienced and recorded by human beings in connection with UFOs, nor may it exclude the mass psychological factors that militate against free discussion of the subject. Many dimensions of the UFO problem are indeed immeasurable.

The UFO mystery is intimately involved with the whole question of how human beings perceive phenomena, and how their perceptions are bio-energetically and bio-psychiatrically distorted. In short, we confront the inherent errors of man's natural philosophy. The idea that UFOs and their technical principles can be adequately dealt with by unaided physicists, engineers and aerospace specialists, supported by ministers of religion – is recklessly naive. The days when we can delude ourselves that we are investigating this subject when we run deferentially to panels of physicists for approval of our findings, or when we try and get Congress to act, are gone forever for all realists. The experts of this earth are experts in what this subject – these phenomena – are not.

UFO phenomena are all around us, unseen. This basic fact has been established by radar and further demonstrated by my pioneering photography. Accurate understanding of such phenomena requires the New Knowledge, in which formal qualifications may well run second, third, or worse to actual participation. Cosmic tides wash strongly against the ivory towers of mechanistic science, and will tumble those ivory towers just as soon as old ways and methods are transmuted by the brilliant young men and women already entering upon careers in science. They are a new and different breed of human, and formal education lags far behind the exceptional powers and capacities that they have brought with them into the world.

Native common sense, activism, unblocked perceptions and a free human being's understanding and acceptance of his own basic life processes, are the primary qualifications for facing UFOs on their own ground. The new young scientists have these faculties and capacities. Their diligent scientific labors in years to come will codify what is merely broached and indicated in this book – in short, they will make it into the science of tomorrow. By summarizing here the tradition-wrecking aspects of UFOs, we can illustrate strikingly the need for a fundamentally new thinking – strong and vital enough to break man out of the straitjacket of the past.

The role of radar in demonstrating the invisibility of many UFOs has already been described, and its role in provoking the phenomena broached for later elaboration. Subsequent to the large-scale development of radar, UFOs also appeared as visible, physical objects, and they have been with modern man ever since. Atomic explosives and atomic power are also interwoven with the nexus of electrical events already cited. The mode of this atomic implication will be clarified when the discovery of orgone energy is elaborated in later chapters.

A comprehensive theoretical approach to UFOs must provide an acceptable explanation, preferably with experimental support, for the visible manifestations as well as for electronic sightings of invisible UFOs. Scientific honesty requires that nothing observed should be evaded. Since UFOs have been seen to appear and disappear on numerous occasions, we must be prepared to deal with transitions of substance from physical to invisible-physical. In my personal experience, I have never once seen a UFO in the normal physical state that did not vanish while I had it under observation.

Since I could not follow these objects with my sense apparatus, my adventures stem largely from following them with my thinking. External apparatus was then used to objectify and verify what my thinking commended to me as being truthful and lawful about their disappearance. By their mode of manifestation, UFOs are inviting us to follow them. That is what they are "saying" when they disappear before our eyes.

The facts as they have unfolded in the UFO field force us to go much farther than understanding even such seemingly incomprehensible happenings as materialization and dematerialization of flying discs. Ignoring the evidence of his scientific instruments has brought into question the methodological and epistemological basis of man's modern science. We have been forced into mass characterological considerations of a dimension and complexity almost as staggering as the UFO phenomena that have thrown these considerations into relief. Errors and inadequacies to date lie less in scientific instruments and materials than in the character structure of those using these tools of investigation. The instrument is worthless in the hands of the man who cannot tolerate biophysically and bio-psychically what it records.

Railing against a dying but still powerful order in science is of no value. Mechanism is there. We are dealing with a definite psychophysical structure in Mechanism that can be understood, but not changed in the mass sense, other than prophylactically. Children raised in accord with New Knowledge principles and emerging into healthy, non-neurotic adulthood, will not easily accept the mechanico-mystical splitting of the human psyche to which prior generations have been forced to succumb. The youth revolution has its roots in the new, healthy wholeness of the coming humanity. This is another of the numerous immeasurable dimensions to the UFO problem.

The mechanistic scientific mind has been oriented for generations to denial of the invisible. This mind has accordingly been unable to approach UFO phenomena that beckon human attention to invisible strata of energy and substance, wherein lie the roots of life. That invisible UFOs suddenly appear to our gaze, and that conversely, visible UFOs suddenly disappear to our gaze, simply tells us that the invisible and the visible are functionally unified. The structural tendency of the mechanistic mind to split apart functional wholes – to shatter to fragments and then bewail the complexity of nature – is well illustrated in this aspect of ufology.

As this book proceeds, the reader will be made aware of what has already been achieved, although not officially accepted, in exploring our invisible-physical borderland and its denizens. My experience has taught me the futility of seeking formal recognition for any of these findings, because such a venture reduces itself always to a hopeless battle against character structure – against modes of reaction and behavior inculcated since infancy. One is forced to choose between continued quiet work and exhaustion in the labyrinths of scientific bureaucracy.

My choice was the former, seeking to demonstrate as conclusively as possible that the invisible is upon us, and to illustrating and illuminating the cognitional impasse in which mechanistic science has landed itself at the time of its greatest triumphs and highest influence. The immeasurable dimensions of the UFO mystery include this constant pressure for change on the old, classical scientific order. Youthful attention turns inevitably toward functionalism – the ability to follow with the mind the perpetual dynamic changes of the living. Acosmo-conception backed by nearly two centuries of continuous academic and cultural support is now under fire through UFOs. We are already a long way from mere ships-from-other-planets.

Tied in with the chronological breakthrough of UFOs came the "foo fighters" of World War II. Scientific non-results in dealing with this phenomenon demonstrate many of the inadequacies of existing cognition and method. To this day the foo fighters have not been satisfactorily explained.

These elusive, seemingly intelligent small objects played luminously around warplanes toward the end of World War II in Europe. Allied reaction was to classify them as a new German weapon under tests, since no casualties resulted from their presence. The Germans thought they were a new Allied invention. After the war, both sides found their evaluations incorrect. Foo fighters belong to no one on earth. During the Korean War, they were seen again.

Harvard University's Dr. Donald Menzel opined that they were reflective eddies created around battle damage to Allied aircraft, but Dr. Menzel didn't carry out his usual comprehensive research. His assertion in his book Flying Saucers,[1] that battle damage to Allied bombers was greater in the final stages of the war is historically insupportable. In the Korean War furthermore, U.S. aircraft suffered little battle damage, due to the lack of consistent enemy air strength.

Foo fighters should be amenable to explanation by the kind of comprehensive UFO theory of which mankind stands in need. Ad hoc theories to cover specific, isolated instances can usually be formulated by specialists when the phenomena appear within the scope of their disciplines, but rarely are such theories useful beyond the specific instances. An approach is needed that can help break down the compartmentalization of knowledge – in itself a consequence of man's propensity to split apart and artificialize phenomena that are functionally unified.

The UFO theory needed should permit an understanding not only of the determinism of foo fighters, but of the functional relationship they bear to all the other invisible objects encountered in ufology and to the more conventional flying discs. Now if we take a highly sensitive Super 8mm camera to 30,000 feet in broad daylight in an airliner, load that camera with standard Ektacolor 160 film and cover the lens with an 18A filter, we have an experimental arrangement that can record evidence. An 18A filter is scientifically designed to absorb all visible light and color. Don't let that worry you. Shoot a few rolls of such color film, with a filter designed to block all color. You'll find the whole gamut of UFOs are out there, and you'll record them in color. Impossible? Don't even discuss the matter until you have acted, until you have been there and done it.

A vista of staggering portent beckons with the use of digital camcorders, similarly fitted with an 18A filter. This equipment was not available when this book was first written over thirty years ago, but now offers those willing to use it a number of amazing possibilities. From a technological standpoint there is no better time in history, for looking straight into this invisible world, than right now.

This is cited to illustrate how, by sometimes going 180 degrees against conventional theories and ideas, objective evidence of UFOs may be acquired. The man to beware of in this field is the narrow scientific specialist. He will be the first man stretched thin by the sheer width of the ufological spectrum. Only activism – work and participation – count in this field.

Solidly verified concomitants to UFOs include interference with electrical systems. Such UFO interaction with man's electrical works date from World War II. The late Harold T. Wilkins reported an early incident involving suspension of an aircraft's electrical system in his Flying Saucers Uncensored, a 1955 opus from Citadel Press of New York. Writes Wilkins on page 209:

" In 1944 an American pilot, flying over the Burma Road, said his plane was held motionless and propellers stopped, while far aloft a mysterious disc appeared to be putting a sort of immobilizing ray on his plane. After this seeming 'inspection' his power came on again, his propellers resumed turning and the mysterious object disappeared into the far blue."

Since that time, the world has seen numerous instances of commercial power failure, suspension of auto and aircraft ignitions, and a variety of magnetic and electrical interference indisputably connected with UFOs. Such happenings must fit into a comprehensive UFO theory, and not be split off for study as discrete phenomena.

Caution towards the narrow specialist is enjoined by the occurrence of such phenomena within the much wider body of UFO phenomena in their totality. In earth science we cannot yet duplicate this ability to paralyze electrical activity. Ford Motor Company tried it unsuccessfully on automobile engines. The theorist concerning such paralytic activity should be asked to account also for kindred and connected UFO phenomena. If we apply this principle practically, we soon learn, practically, that much of our so-called "hard" knowledge is illusion. Here again, it is the orgone energy – the ether – with its demonstrable antagonism to electromagnetic energy, that provides the technical break-in.

Every scientific specialist faces a stupefying spectrum of phenomena connected with UFOs that lies outside his discipline, in areas in which he is not qualified technically even to hold an opinion. That is why Dr. Wilhelm Reich was right to say that there are no "authorities" and no "experts" now that cosmic science – the New Knowledge – is being literally forced upon us. The rational approach is a sharpened awareness of the inadequacies of mechanistic science in dealing with Cosmic phenomena. We are all brethren in ignorance, facing immeasurable new dimensions.

Hostility on the part of certain UFOs is another factor that must find its place in a comprehensive UFO theory. Establishment ufology has a blind spot here. This aspect of the UFO problem has been steadily resisted, despite the evidence that aircraft have been destroyed in the air – and sometimes kidnapped complete with crew. In my 1958 book *They Live in the Sky*, the affidavit of a veteran French pilot, M. Pierre Perry,[2] was presented. Perry recounted a shocking incident he had observed from the ground in the wilds of Arizona in 1943.

A USAF aircraft with two occupants was deliberately destroyed by balloon shaped UFOs. The bailed-out crewmen had their parachutes set on fire by heat rays from the UFOs. The unfortunate victims of these weird entities from space were crushed to death by their fall to the ground. Few indeed are the ufologists who will look directly at such happenings and see them for what they are. So-called "objective" investigators have preferred to disappear into the mist of wishful thinking.

In the same book, I presented another sworn case from Paris, Illinois, wherein a USAF jet fighter was observed from the ground by Mr. Eugene Metcalfe to be abducted in flight by a large, bell-shaped craft of unknown origin. The jet was simply swallowed into the underside of the hostile vehicle. This was one more instance of seriously unethical acts by entities from space, but the establishment in ufology declines to see such acts in all their clean clarity.

There was also the abduction over Lake Superior on 23 November 1953 of a USAF F-89 jet fighter piloted by Lt. Felix Moncla Jr., whose aircraft was vectored to a UFO by ground control intercept at Kinross AFB, near Sault Sainte Marie, Michigan. Moncla's fighter merged on the radar screen with the UFO, the two objects becoming one large blip 70 miles from Keweenaw Point. Moncla and his radar observer, Lt. R. R. Wilson, were never seen again, their aircraft was never found, no wreckage was recovered, and the pursued UFO also disappeared.

Major Donald Keyhoe gave a full account of this baleful incident on pages 13-23 of his *Flying Saucer Conspiracy*. Despite this and many other similar incidents, until the end of his tenure as Director of the National Investigations Committee on Aerial Phenomena (NICAP), Major Keyhoe believed there was no convincing evidence of UFO hostility. Lts. Moncla and Wilson had vanished from human ken, complete with plane, but even this is deemed unconvincing evidence of hostility. Again we find, in a new way and in another facet of the UFO subject, that same evasion of the essential that has kept ufology tied to the skirts of the mechanistic world conception.

There have been numerous cases involving hostility on the ground in encounters between humans and a variety of queer entities that have dismounted from spacecraft of various kinds. Humans have been attacked and clawed, their abduction attempted, and others have been knocked senseless by various ray weapons possessed by the intruders. These incidents have occurred year in and year out, in areas as widely separated as South America and Scandinavia, and have been verified by responsible investigators.

Some of these incidents will arise later in this book, in context with new findings, but the serious student of ufology desiring a steady flow of such information can do no better than subscribe to the Flying Saucer Review from England.[3] FSR presents the stories of these vital incidents after investigation by its qualified representatives in foreign countries. The publication is produced avocationally by a group of scientists, engineers and physicians whose qualifications are beyond reproach.

Self-styled skeptics avoid looking at these unsavory, unwelcome and disturbing facts. Weak jokes about "little green men" constitute the maximum effort mounted by the media of the western world with regard to these epochal encounters. Any intelligent, alert, unblocked and discriminating individual can satisfy himself quickly concerning the validity and the increasing incidence of these landings and encounters. They are no joking matter.

The hostility of certain visitants must be woven into a comprehensive UFO theory. Since the incidents continue with the years, and apparently began with the war period, we should expect a functional connection to exist with the other complex and seemingly impenetrable aspects of UFOs. Once more we may note that the narrow scientific specialist in a technical discipline can bring little to bear on this serious and far-reaching problem of ethics and behavior. Conventional psychiatrists and psychologists who might assist here cannot deal with radar or electromagnetic interference. The discovery of the orgone energy – alone among all events of the past several centuries – gives functional access to all these riddles. A new epoch has opened.

The obverse aspect of hostility is the reluctance of most UFOs to make contact with humans. Most UFOs are elusive. In most cases, the objects make off at high speed from the vicinity of human observers, from aircraft or from happenstance encounters with humans. This instant readiness to conceal themselves from human beings is well established, and must take its place in any comprehensive UFO theory. There must be a reasonable basis for saying why these things happen so frequently as to be among the basic characteristics of most UFOs.

The immeasurable dimensions of the UFO problem may be seen also in a worldwide outbreak of psychism. Associations between psychic phenomena and UFO phenomena are intimate, broad and profound. Dozens of books have already been written about UFOs on a psychic or quasi-psychic basis, and there is a potent mystico-religious overtone to all UFO affairs that may not be ignored – any more than we may ignore the evidence of radar.

The coming of the flying saucers has been the subject of numerous sermons by ordained ministers, and in the more modern churches unballasted by tradition, UFO lecturers are sometimes invited to speak at Sunday services. These speakers are free to ventilate the metaphysical aspects of the subject as they see fit. The heads of large aerospace concerns and distinguished men of science may be observed often in the congregations. Some of the braver ones have even come to my talks.

A sociological fact of life in today's world is that millions of people have been led to absorb themselves in the metaphysical aspects of UFOs, whether or not they are possessors of formal scientific training. There is no manifestation in human record that comparably straddles science and religion, straining each division of thought at its foundations. The purportedly objective approaches to UFOs that ignore these staggering sociological facts must be characterized as fraudulent.

Our comprehensive UFO theory must look at this situation squarely, and establish how it has arisen, and why. Ignoring the irruption of psychism that is coincident with the modern advent of UFOs is tantamount to ignoring the radar evidence, the photographic evidence or any other kind of evidence. There must be a place in our theoretical mosaic for these undeniable psychic, mystic, occult, spiritualistic and religious aspects of the UFO subject because they not only exist, but also are extending their influence continually into the vacuum left by scientific abdication.

The socially-endorsed posture of the mechanistic scientist, that these manifestations are "mysticism" beyond ken and therefore excluded from scientific attention, is not only cowardly, but mindlessly evasive. By giving these manifestations a name – whatever that name may happen to be – the mechanistic investigator deludes himself that he has explained what is happening. The deep stirrings evoked in millions of humans by UFOs have taken place at the dawn of mankind's Cosmic Age. Conceiving of this additional mighty coincidence as an accident is the hallmark of a simpleton.

Our comprehensive theory of UFOs must face these socio-religious phenomena, and the undeniable impact of UFOs on the inner life of Man. Honest investigation of UFOs and corollary observations of 20th century life, will convince us that spiritual forces are at work on, in and through the human being in a decisive fashion, and with definite and comprehensible earthly goals. Right now, this is yet another immeasurable dimension to the UFO problem, but it is there and must be faced.

Contact and communication with UFOs have been derided not only by establishment-type ufology, but also by a lamentable coterie of qualified scientists. A comprehensive UFO theory must provide an understanding of the psychic encounters with UFOs that far outnumber physical encounters. Processes beyond the reach of official science have obviously been brought to bear on human beings by the aliens. These encounters are as indubitable as the physical landings and incidents, even if inaccessible to mechanistic method. New ways must be found and opened. UFOs are too important to be left to mechanists.

A viable UFO theory must account for the lack of communication through orthodox, electrical communication methods. There must be sound reasons why advanced UFO intelligences do not utilize our communications systems to signal us in some way, and we must have at least a theoretical guideline to the means of communication they do employ. Psychic communication has been extensively employed in many variants, and the only rational attitude is again to face these facts and let them lead us where they will. Our existing radio receivers have got us nowhere. Our psychic receivers have brought in plenty.

Man is irrational when he makes Herculean efforts to become a cosmic voyager on the one hand, and denigrates communication with alien intelligences on the other. This is the situation today in the official attitudes toward communication with space. Extraterrestrial life remains the largest challenge to mankind growing out of UFO phenomena, and mechanistic biology, rooted in sterile chemistry and physics, has practically no chance of cosmic survival. UFO data already on record sharply illustrate its inadequacies.

The fundamental characteristic of everything that is alive, as opposed to what is inert, is pulsation. To official biology, pulsation remains inaccessible, inscrutable – a mystery as vast as the UFOs themselves. Prompted by observed pulsations of UFOs sighted at high altitudes by its pilots, the USAF once broached the idea of "space animals" in a public release dated 27 April 1949, stating that the objects appeared to behave more like animals than anything else.

The airmen who observed these aeroforms, and expressed the feeling that they were living organisms, were closer to the truth than scientists in the grandiose discipline of exobiology have yet come. A comprehensive UFO theory must incorporate living aeroforms within its structure, for such living creatures do exist, have been extensively photographed by myself and others and are now being recorded inadvertently on NASA videotapes in the space environment. NASA's exobiologists literally do not know what they are looking at, so intent have they been on establishing the sterility of the moon. Some of these creatures are monstrous in size.

Many persons have theorized that UFOs are alive, including the late Kenneth Arnold, who put the term "flying saucer" into the language. I happen to be the earthman who first photographed these life forms extensively, proving their existence and simultaneously penetrating their bedrock involvement in the UFO mystery. A comprehensive UFO theory must include them, with all their revolutionary portent for the new life sciences, and their power to shatter some of mechanistic man's cardinal illusions about the origins of the earth.

UFO propulsion should have received instant and unrelenting scrutiny by world science, for it is beyond doubt that a new mode of propulsion is involved. Existing instrumentation and knowledge admit us only to fringe physical effects of this power. A different kind of power, a heretofore unknown or unsuspected mutation of energy is involved. UFOs exhibit mastery of gravity, and can attain velocities and execute maneuvers far beyond the farthest reach of mechanistic science or its most optimistic projections. The UFOs are in the here-now with all this. Air-supported craft are destined for the dustbin.

Man has nevertheless been content to waste his scientific substance in chemical power plants of monstrous inefficiency, with precarious control of space ventures exercised only through the precise cooperation of battalions of highly skilled technicians and specialists, who crouch convulsed with fear at launch time lest their plaything fall over and blow up on the pad. All this tragicomedy goes on while space vehicles are already present in the atmosphere, and beyond, that eclipse our ashcan spacecraft as a Cadillac surpasses Ben Hur's chariot.

Brave astronauts riding these clumsy, manmade contrivances into space have sighted and photographed vehicles otherwise propelled and controlled, yet the focus of aerospace activity stays on rockets and kindred devices as though there were something to be feared in the energy system that is propelling the UFOs. This fear and avoidance of the new power by human beings in the most relevant of all fields, has to be explained by our comprehensive UFO theory. Again, we will find as we assemble this theory – with both thought and experiment – that it is the orgone energy that makes such explanations possible.

Our theory must concern itself not alone with UFOs and their scientific determinism, but with the thorny, prickly question of irrational human attitudes toward the phenomena. Our concern here cannot be confined to the layman or to the nut. A solid theoretical approach should allow us to understand why scientifically trained individuals have not only avoided this subject, but in many cases have become active and even ardent agents of the new obscurantism.

There are the visible manifestations, stunning in themselves, with their irrefutable invisible aspects; there is the connection between the electromagnetic nexus of the Second World War and the modern, worldwide advent of the phenomena; there are objects that are seemingly spacecraft and objects that are obviously biological; there is the paradox of hostile visitants with heat rays, murdering humans in remote areas, and the general avoidance of contact with humans; there are the psychic, metaphysical and occult aspects of the mystery; there are the great enigmas of communication with the piloting intelligences; there is the riddle of pathological evasion of the UFO problem by human beings deemed responsible in all other normal ways; there is the new mutation of energy that is the key to unraveling the tangle – the swaddling cry of a new Life-positive science that the doomed priests of the old order seek to strangle at birth. There is the conceptual dead end at which mechanistic science finds itself at its moment of greatest glory and influence; and there is the glittering promise of the New Knowledge arising from the ruins of the old.

When our comprehensive theory is assembled, we should be able to understand why men, enlightened and educated to the best formal standards, turn away in the clutch from science in favor of scientism. We should be able to understand why those scientists who accept that UFOs are ships from outer space, appear structurally incapable of dealing with the inevitable, consequent question of contact with the piloting intelligences. A comprehensive UFO theory must elucidate, and thereby prepare for the eradication of, these obstructive human dilemmas that affect all people to some degree.

These are just the primary aspects of the greatest mystery in human history – a mystery that requires a changed human being for its eventual penetration. Consider the range, scope and depth of the UFO problem without prejudice. Ask yourself if the so-called extraterrestrial hypothesis, or ETH, is adequate. Ask yourself if your own present thinking and manner of viewing creation is adequate to cope with these presently immeasurable dimensions.

The towering theoretical problem involving UFOs is a challenge to the best that is in man. The Little Man that is in all of us, usurper and suppressor of all that is great and godly in every one of us, wants us to hang on like drowning people to that seedy ETH. If we catch the sweeping magnitude of the mystery, if we let its grandeur and cosmic power live in us, then there will be room no more for our Little Man. He is the one who whispers to every one of us that we can do nothing great – that all people are Little.

The Little Man wants you to recoil from the vastness of this subject. When the problems are presented as they have been here, already interwoven with each other so that we can see the folly of segmental, indecisive approaches, we can already see why there have been no government announcements about UFOs that have any meaning whatever. Only an idiot – a Little Man – can expect or feel the need for such announcements. The subject is too vast, too far-reaching, and too radical in the primary meaning of that term, "root" – to permit quick assurances to the Little Man. This subject is big.[4]

There are functional connections between all the ramified and seemingly irreconcilable factors thus far outlined. Functionalism gives us a new beginning and a new way to proceed in the future. UFOs are manifold and multiform phenomena that are at once ancient and ultramodern, containing within themselves a range of physical, biological, biosocial, biopsychic and bio-economic principles new to mankind. UFOs are the space age bearers of the New Knowledge, knowledge that has the power to renew human life and culture.

Man reached this cosmic rendezvous with a fragmented and essentially contrived mode of scientific cognition, and with his intuitive intellect beaten back into a corner like a whipped dog. Man must make himself whole and healthy if he is to plunge farther into the cosmos, and have commerce with the beings who are even now all around his planet Earth. He will need to keep a firm hold of the best in what he has learned, but his greatest need is to get a firm grasp of the new. What is worthless and devoid of value in his Cosmic Age, no matter how old or honored by tradition, can then be allowed to slide into the limbo.

For the individual, a prime task is to push his usurping Little Man off his own inner throne and put the King in his place. Dr. Franklin Thomas taught me this, and thereby opened all that followed. Every human being can do likewise.

By turning a substantial portion of my life energy and my earnings for many years into the pursuit of the New Knowledge; by taking advantage of opportunities that came to me to learn from several magnificent human beings; and above all by going out and participating as an innovating experimenter – daring to do – I believe that I have reached the end of the beginning of the UFO mystery. My Little Man would have convinced me, years ago, that an ordinary man could never do such a thing alone and unaided.

I believe I can formulate for you the comprehensive theory that seemed so frighteningly complex only a few pages ago. What it took me painstaking years, grinding struggle and immeasurable sadness to find out, you can learn in the short space of this book. Stand by my elbow and follow my story as it happened. Grapple with the UFOs at first hand as I did, and share my great adventure.

NOTES

1. Published by Harvard University Press, Cambridge, Mass., 1953.
2. The author met M. Perry personally, and questioned him in detail.
3. Ufology's best buy. Flying Saucer Review, FSR Publications, Ltd., PO Box 585, Rickmansworth, WD3 1YJ, England. Website: www.fsr.org.uk.
4. For further verification of UFO reality see Col. Philip Corso's book, *The Day After Roswell* (1997) and the following links:
http://www.bibliotecapleyades.net/exopolitica/esp_exopolitics_ZZO.htm
https://en.wikipedia.org/wiki/Philip_J._Corso
https://en.wikipedia.org/wiki/The_Day_After_Roswell
See also *The Flying Saucers Are Real*, by Donald Keyhoe; *Camouflage Through Limited Disclosure: Deconstructing a Cover-Up of the Extraterrestrial Presence*, by Randy Koppang; *UFOs Do Not Exist: The Greatest Lie that Enveloped the World*, by Dr. Roger Leir (Foreword by astronaut Dr. Edgar Mitchell); *UFOs and the National Security State: Chronology of a Coverup, 1941-1973*, by Richard M. Dolan (Foreword by Jacques Vallee); and the following links concerning The Battle of Los Angeles, California; invasion of 24 February, 1942:
https://www.youtube.com/watch?v=EhjkMoWLE_Y
http://www.rense.com/ufo/battleofla.htm

Chapter 4

ETHERIC RAIN ENGINEERING
Operations Summary

Over a developmental period exceeding 30 years, etheric rain engineering has been utilized in numerous successful operations. Beginning with ponderous, fixed-base operations employing arrays of up to 150 water-grounded tubes, this art and science has continuously evolved. Advances have been both technical and theoretical. By the year 2002, the huge, static, water-grounded arrays have been replaced by a single, hollow tube of sensitive construction. These tubes are carried aboard helicopters or light aircraft, and do not employ chemicals, electric power, or electromagnetic radiation in any form. From inception until today, action has depended on tapping the latent, native power of the etheric continuum.

Effort has not been expended in reconciling this new technology with scientific conceptions and theories that exclude, a priori, the existence of the ether. On the contrary, every effort was made to keep this work independent from classical theories and notions already obsoleted by the practical results that have been obtained over the past three decades. Instead of seeking the approval and sanction of formal scientific sources, objective physical results were allowed to guide and discipline development. This unorthodox approach ruled out any drifting to mysticism. ONLY RESULTS COUNT is the motto that ruled.

Typical of the wide range of etheric engineering operations were the following, documented projects:

OPERATION KOOLER was carried out in September of 1971. A devastating, entrenched heat wave (106F) was crippling Los Angeles and Southern California. Electric power distribution was beginning to break down, schools were closing and wide popular distress grew worse as the crisis conditions continued without surcease. Contra-forecast, KOOLER brought regional temperatures down by 31 degrees Fahrenheit in less than three days, ending with light rain.

SYCAMORE CANYON FIRE 1977. A commercial rain engineering project was being carried out in July of 1977 for the Eastern Utah Cattlemen's Association. Fixed-base weather engineering units were employed in both southern California and Utah, and the project was successfully concluding when a severe heat wave hit southern California. This prompted the engineers to attempt a repetition of the highly successful KOOLER operation mentioned above. A diversion from the Operation Reviver format used for Utah, was accordingly filed telegraphically with the U.S. Federal authorities (NOAA). The relevant telegram survives to this day. Immediate adjustments were made in the desert-based equipment in California. The engineers were not aware, in driving to the desert base, that a ferocious fire had erupted in the Sycamore Canyon behind Santa Barbara city. Flames were being driven downhill by scorching, 55 knot winds. The flames had reached to within 800 yards of downtown in the beautiful city, which appeared lost. Then, at 12:45 am, approximately 30 minutes after our engineers had completed aligning all equipment on the magnetic west vector out of Thousand Palms Oasis, the winds suddenly reversed, with a precipitous temperature drop from 90F to 71F. The fire was blown back up the hill, saving the city of Santa Barbara. The magnetic west vector out of Thousand Palms goes right through Santa Barbara. The fire was brought under control after this (unwitting) intervention of etheric weather engineering into a potentially disastrous situation.

HIGH SEAS OPERATIONS 1967-1992. These operations were carried out aboard ocean-going merchant ships of the U.S. Merchant Marine, in which Mr. Trevor James Constable served as a Radio Electronics Officer. His professional responsibilities included ship's radar installations. For the first time, it became possible to operate etheric rain engineering equipment in a mobile format, in a pristine environment, and with the unrivaled documenting and observational properties of modern radars Progress was immediate, particularly in understanding the practical workings of the etheric continuum and its laws, in all kinds of climates and environments. . Permanent assignment in 1978 to the SS "Maui," flagship of the Matson Navigation Company's fleet, further stabilized these investigations of the maritime mobile format. Mr. Constable made more than 300 crossings of the eastern North Pacific before his retirement from the sea in 1992.

Numerous etheric rain engineering devices were designed, constructed and tested during this period. Only those designs that produced practical results were further developed. Rain was consistently engineered from SS "Maui" within major high-pressure cells, some of them covering a million square miles. Results were comprehensively documented on time-lapse videotape. A commercially-released video, *Etheric Rain Engineering On The High Seas*, summarizes this pioneering work in the maritime mobile use of etheric rain engineering.

In the course of his 14-year tenure as Radio Electronics Officer of SS "Maui," Mr. Constable conducted literally thousands of rain engineering experiments of all kinds. He brought down literally billions of tons of water on the earth with these explorations of nature's great laws. The great majority of these experiments were objectively successful, and contributed to the development of an invaluable OPERATING ACUMEN.

OPERATION PINCER II, July 1986. An engineering drawing was filed in advance of this project with the U.S. Federal authorities (NOAA), detailing the procedure that would engineer rain into Los Angeles in July 1986. Los Angeles is "*statistically rainless*" in July. Etheric rain engineering was being pitted against a century of weather records. PINCER II demonstrated the anomalous diversion – 240 miles out of its normal path into Arizona – of rain that made July 86 the wettest July in 100 years. The attendant, anomalous thunder and lightning storm left 300,000 Los Angeles residents without electric power. Entire operation was documented by US government radar fax maps and storm video.

OPERATION CLINCHER – Smog Dispersal, May-November 1990. Federally-filed in advance with NOAA, Clincher had as its announced target, a twenty percent *seasonal reduction* of smog. This was to be across-the-board, for the huge four-county Air Quality Management District, the largest, filthiest smog region in America. This six-month operation utilized 14 etheric vortex generators that had been technically developed aboard SS "Maui." No chemicals, no electromagnetic radiation. Smog took the worst knock, in every category, since smog began. Reductions of over 60 percent in some of the worst areas. Zero smog in several. Overall seasonal smog reduction: 24 PERCENT. Comprehensively documented with AQMD statistical releases.

OPERATION TANGO, Singapore, July 1988. Dry season rain engineering. TANGO, like CLINCHER against Los Angeles smog, was sponsored and financed by a young Singapore entrepreneur, Mr. George K.C. Wuu – one time President and Chief Executive Officer of Etheric Rain Engineering Pte. Ltd. TANGO utilized a 46-foot Hatteras cabin cruiser offshore, coordinated with an Apache etheric vortex generator at Loyang, on Singapore Island. TANGO raised 30,000 square kilometers of rain over Singapore, Malaysia and northern Indonesia. TANGO was documented with Singapore government radar maps and mobile video.

In addition to TANGO during this Singapore visit, Mr. Wuu and Mr. Constable carried out operational experiments with a speedboat and a rotating "Termite" device. These operations were extremely enlightening as to the special etheric properties of tropical weather.

OPERATION PIONEER – Melaka, Malaysia, 1991 Dry Season. Fixed base operations, with ancillary gun car operations on regional roads. PIONEER brought in 38 measurable rains in 57 dry season days, for a total of 327 mm. This is 75 percent of an entire year's normal Melaka rainfall. Melaka is the driest state in mainland Malaysia. Government radar nevertheless documents unerringly that Melaka was *frequently the only state in malaysia getting rain* during the entire period of PIONEER.

OPERATION RED BARON, Hawaii, 1994-96 – This was the initiating, pioneer series of airborne etheric rain engineering operations. Rain was successfully engineered from the air, using specially designed "Bull" translators, right from the first flight made for this purpose. Conclusively established the power and efficiency of airborne rain engineering, as superior to any previous modality. Methods developed in 14 years of maritime mobile work aboard the SS "Maui" proved much more effective from an aircraft. Documented by video and via third party media reports. Incorporated in special demonstration video for government ministers and senior officials. Has been available from ERE Singapore on letterhead request.

OPERATION SEGAMAT, Malaysia, October 1997. AEREO – Airborne Etheric Rain Engineering Operations – proves effective in a series of exploratory operations in rural Malaysia. These rain flights exploited the advantages of isolated airstrips without traffic control. The program culminated in Segamat, in central Malaysia, where a lengthy dry spell had droughted the region. The Segamat Country Club had its own airstrip. On 14 October 1997, two short, successive rain engineering flights were scheduled, right off the runway directly to the east. Each flight was to be for no more than 20 minutes at minimum safe speed. The second flight had to be curtailed to 9 minutes because of imminent heavy weather. Eight hours of rain into droughted Segamat ensued, which extended some 60 km to the north and west. Documented with radar fax maps, video and photography. (See the 9-exposure photo recording of this operation on page 61.)

OPERATION K.L., Malaysia, April 1998. The Kuala Lumpur region was in a four and a half months period of drought. Water rationing was in force. 28 hours after Mr. Constable set foot in Malaysia and made an immediate AEREO flight, the region was deluged. In the subsequent week, a helicopter was used for the first time in AEREO. Slower, controlled flights in the 40-60 mph speed range, proved super-effective. Coordination of fixed-wing and helo AEREO flights successfully pioneered, revealing a rich potential. Nocturnal rains were also engineered into Kuala Lumpur by means of a fixed-base P-gun installation on the hotel roof. These rains were accompanied by massive thunder and lightning, on the scale of a bombing attack. Entrenched drought in Kuala Lumpur was countered by natural means, without chemicals or radiation.

OPERATION PROFITA – People's Republic of China. September 1999. Sponsored by a Hong Kong charitable foundation. Supervised by Mr. Chen Feng, President and CEO of Hainan Airlines, this was a demonstration operation. A single, 20 minute helicopter flight to the east, equipped with a single P-gun, produces rain over Hainan Island, contra-forecast and as promised, within 12 hours. Extensive rain resulted in the cancellation of the second demonstration flight, scheduled for the following day. It was still raining from the first pass. This was the first airborne etheric rain engineering operation ever conducted in the People's Republic of China. Documented and presented in a brochure by PRC meteorologists and Hainan Airlines technicians.

These operations show the lengthy, varied operational history behind etheric rain engineering. Every modality successfully used remains valid in its way to this day. Technical progress, plus ever-increasing understanding of the etheric continuum and its laws, simply obsoleted the various translators devised in developing and refining this technology. Airborne etheric rain engineering stands as the salient success as we enter a new century, Etheric technology will shape the human future, in ways barely imaginable today.

Chapter 5

THE MARITIME MOBILE EXPERIENCE

Updated October 9, 2008

Trevor James Constable's permanent assignment to SS Maui, on her maiden voyage in 1978, provided an unprecedented opportunity to further develop etheric rain engineering. During the previous ten years in his sea career as a Radio Electronics Officer, Mr. Constable had been able to organize only sporadic, hobby-type experiments, because all his seagoing assignments were temporary and usually for one voyage only. What he was able to do during this hectic period was pregnant with promise. Operating in the pristine environment of the high seas, with radar to objectify results engineered far beyond direct observation, was completely different to fixed-base experiments in the difficult, semi-desert environments of southern California. Operations were far more readily successful. Transfers to a different vessel after every voyage prevented any organic development of this research.

With permanent assignment to the Maui, etheric rain engineering experiments could now be steadily and systematically conducted aboard a fast, brand-new vessel, equipped with the latest radars. The ship itself was an official station of the World Meteorological Organization. The ship's radars were among Mr. Constable's professional responsibilities as Radio Electronics Officer The initial experiments aboard the Maui were authorized by Commodore C.C. Wright Jr., senior master of the Matson Navigation Company, late in 1978. His successor, Commodore Kenneth R. Orcutt USNR, extended this authorization. The flying bridge of this 720-foot long, 23-knot commercial vessel, flagship of the Matson fleet, became a mobile base for etheric rain engineering experiments. Maritime mobile techniques could now be explored and developed, especially the important engineering component of *velocity*. This could now be studied in operations conducted on gyro-stabilized courses. 3cm and 10cm radars kept operational experiments firmly anchored to, and guided by, results.

Matson Lines' container ship SS Maui is seen here off Diamond Head, Honolulu, on her maiden voyage in May of 1978. Trevor James Constable was permanently assigned to the ship as Radio Electronics Officer after a chain of incredible coincidences made the appointment available to him. From 1979 until 1992, the SS Maui served as a traveling laboratory for TJC's maritime weather engineering experiments. The 720-foot long ship was stable and fast (22 knots), and TJC mounted his equipment on her flying bridge, a perfect experimental setup. Even Matson Lines' Chairman of the Board took an interest in the weather work abouard the Matson Lines flagship. The ship became famous in the eastern North Pacific as the "rain ship."

There were no textbooks to consult on this radically new kind of engineering. Results provided a discipline all its own. If a stratagem or a new-design of translator did not produce objective results after a searching trial period, that approach was discarded or modified. Since the radars were a decisive adjunct to the eyes, that which did not produce results was considered a failure.

The theoretical descriptions of the earth's ether economy that are provided in Dr. Guenther Wachsmuth's classic *Etheric Formative Forces in the Cosmos, Earth and Man*, could now be tested empirically, in the pristine high seas environment. The Maui work began with this, a task of synthesis: Dr. Wachsmuth's theoretical scenario should prove confirmable via Dr. Wilhelm Reich's cloudbuster. In the initial investigations, large-diameter (8-inch) water-grounded tubes were employed. This was the cloudbuster in its simplest form. There were no complicated variants when this work was initiated.

In Wachsmuth's sketches of the earth's ether economy, a flow of ether in the temperate zones of the earth moves from west to east. Using the crude cloudbuster and testing this concept on the speeding ship with one of Dr. Reich's basic stratagems, it proved readily possible to dam up this west to east flow in fair weather, so that rain showers would develop "out of nothing" and come right over the ship. Some of the most compelling rain engineering video ever recorded shows this happening from zero, with an immense black band of moisture progressively developing for the time-lapse video camera. Several examples of this action are contained in the publicly released video, *Etheric Rain Engineering on the High Seas*.

Similar investigations verified the existence of the east-to-west etheric flow that prevails in tropical latitudes, with abundant video documentation in time-lapse. Verified as well were the north-to-south winter etheric flow of the northern hemisphere winter, and the south-to-north etheric flow of the northern hemisphere summer. Dr. Wachsmuth was right. So was Dr. Reich. The daily involvement of the chemical ether's "breathing" cycle, in and out of the earth, was verified over the years to be an intimate part of the earth's ether economy. The "inhalation" of the chemical ether into the earth in the evening hours, for example, is a contractive function that enhances rain engineering activity. Spectacular nocturnal rains can be engineered by exploiting the existence of the earth's in-breathing, and especially by the engineer *understanding what is going on*, instead of blindly pushing buttons.

Vessel velocity in certain formats, greatly amplified and facilitated, rain engineering techniques through which results were tortuously obtained ashore from fixed bases in dry climates. In other formats, vessel velocity nullified normally dependable rain engineering techniques. Simple etheric vortex generators were designed, tested and proved, tearing great blue holes to the zenith and squashing huge cloudbanks down around the horizon like a gigantic doughnut. Compelling video proved that such simple etheric vortex generators worked effectively from six decks down in the vessel's storeroom, surrounded by not less than half an inch of marine steel. Radars provided continuous electronic evidence and verification of all these etheric rain engineering results in the maritime mobile mode. Time-lapse videotape and other supporting video, provided irrefutable evidence of the physical existence of the ether. When Sir Oliver Lodge said a century ago, that "the ether is a physical thing," he was right on the mark. Sir Oliver did not know how to get hold of the ether. On the Maui, Mr. Constable found out how that is done.

In more than 300 Pacific crossings, literally thousands of rain engineering operations were conducted. Billions of tons of water were brought down out of the atmosphere. When it became essential because of shipboard requirements to completely abandon water grounding – the original operational format of Dr. Wilhelm Reich – successful designs were contrived, tested and proved that freed etheric rain engineering from dependence on water grounding. All along, water grounding had been a severe handicap dating from pioneer discoveries. Completely new types of translators were now feasible, and led to the present-day sensitively constructed single hollow tube types, with no chemicals or electric power.

A supremely important adjunct to the 14 years of experimental work on the Maui was the intangible asset of operating acumen. This corresponds to the young doctor fresh from medical school, concerning whom it is reasonably possible that he will not kill anyone through his inexperience. When he is 50 years old, he is far more effective because of his medical acumen, acquired in years of practicing his healing art. In the same way, many years of rain engineering operations develop the special acumen for carrying on this art successfully. In etheric rain engineering, both the intent and the experience of the operator will influence success.

The following excerpt is taken from the updated version of *Loom of the Future*, an interview book presenting an encounter between Thomas J. Brown, the then-director of Borderland Sciences Research Foundation, and Trevor James Constable. Mr. Constable states:

> Many times I stood on the Maui's bridge with Commodore Orcutt, as the ship raced along before a following wind and sea. This resembled low flying. The Commodore's genius mind would muse on getting my earthbound gear airborne: 'If you could develop something mountable on a light aircraft, I think you'd get superior results.' He dilated on this theme repeatedly through the years. As a skilled pilot and flight instructor, he was extremely air-minded. The Commodore never told me what to design or develop. His way was to indicate. Action was up to me. A computer specialist of immense ability, his mind functioned by just popping up vital information – like a computer.

> These discussions kept me air-minded. Airborne etheric rain engineering would be our crowning achievement, but the rotating devices we were using successfully on shipboard could not be legally and safely mounted on a light aircraft. Developments had to wait until after my retirement from the sea in 1992. A breakthrough came in January of 1994 during anti-smog experiments in Las Vegas. A small, simple translator resulted that could be safely attached to a light aircraft. Precise dimensions and careful manufacture produced a rugged mobile translator. Would this simple structure, with no chemicals or electric power of any kind, and no spinning component, actually produce results through airborne employment?
>
> Initial tests in Hawaii revealed a stunning potential. A new epoch had opened. All the toil, travail and expense since 1968, appeared now as an essential preparation for the airborne mode. My own experiences, especially the Maui years, had endowed me with a crucial asset: operating acumen. I knew how to apply this vital new tool, what to look for and expect, and how to begin controlling the extensive results. Since the first tests in Hawaii, tests and actual operations in Malaysia, Hawaii and People's Republic of China have verified the dependable effectiveness of of airborne etheric rain engineering in equatoria. Further research will render these methods equally effective in the world temperate zones. Maritime mobile provings there have been conclusive.

That is the linkage of the fruitful Maui years to the airborne etheric rain engineering operations of today. Maritime mobile operations provided the foundation on which airborne etheric rain operations can prove to be one of mankind's greatest blessings, removing the curse of drought from the human future.

Updated October 9, 2008

Chapter 6

AIRBORNE ETHERIC RAIN ENGINEERING

My previous book, *Loom of the Future*, summarized 25 years of research into non-aerial based weather engineering equipment and research. Each development led functionally to the next, and the shape of things to come was obvious – the long foreseen use of aircraft. It became clear by 1990 that this would be the definitive phase of etheric rain engineering development.

Many times, I stood on the bridge of the SS Maui with Commodore Orcutt, as the ship raced along before a following wind and sea. This resembled the act of low flying. The Commodore's genius mind would muse on getting my seaborne equipment airborne. "If you could develop something mountable on a light aircraft, I think you'd get superior result." He dilated on this theme often through the years. As a skilled pilot and flight instructor, he was extremely air-minded. The Commordore never told me what to design or develop. His way was to indicate. Action was up to me. A computer specialist of immense ability, his mind functioned by just popping up vital information – like a computer.

These discussions kept me air-minded. Airborne etheric rain engineering would be our crowning achievement, but the rotating devices in successful use aboard the Maui could not be safely or legally mounted on a light aircraft. Development had to wait until after my retirement from the sea in 1992. Separation from the ship slowed things, but I kept going. A breakthrough came in 1994 during anti-smog experiments in Las Vegas, Nevada. A rugged, simple translator resulted, suitable for safe attachment to a light aircraft or helicoptor. Would this carefully designed but empty tube, without chemicals or electric power in any form, actually engineer rain through airborne use? In the crucible of practical experiment, we answered that vital question.

Airborne tests in Hawaii revealed a stunning potential. All the toil, travail and expense since 1968 appeared now as essential preparation for the airborne mode. My own long experience had endowed me with a crucial asset: operating acumen. I knew how to apply this magical new tool, what to look for and expect, and how to begin controlling the results. Subsequent operations in Malasia, Hawaii and the People's Republic of China have verified the dependable effectiveness of airborne etheric rain engineering operations in equatorial. Similar effectiveness can be anticipated in temperate climates, where maritime mobile operations have been comprehensively effective and documented. The technology was recently controlled by Etheric Rain Engineering Pte. Ltd., a Singapore company formed originally by my trusted friend, George K. C. Wuu, who was its Chairman and CEO.

REVOLUTIONARY DEVELOPMENT

ETHERIC RAIN ENGINEERING GETS AIRBORNE

Effective airborne translators were designed in 1994, initiating a revolutionary advance in etheric rain engineering. These translators, now known as "P-guns," are less than four feet long, with small cross section and light weight. Skilled pilots reported no change in flight characteristics from wing strut installations. Newer variants are carried inside the aircraft or helicopter, obviating aerodynamic problems and permitting easy, in-flight deactivation for enhanced safety.

Airborne tests in Hawaii and Malaysia proved that mobile rain engineering techniques developed in 14 years of high seas experiments, are far more effective in the airborne mode. The much "longer" horizon from the air, the adjustable speed and infinite heading control of aircraft, produced objective rain responses with unprecedented rapidity. Targeting control of the rainfall drop also became feasible. Etherically-engineered rainfall triggered by airborne operations, frequently lasted for many hours. Results became far more certain than ever before.

No chemicals, batteries, electric power or electromagnetic radiation in any form are utilized. Airborne etheric rain engineering is ENVIRONMENTALLY PURE.

When aircraft use became feasible, etheric rain engineering developed a powerful new dimension. Vast droughted territories in equatoria may now be readily accessed. Pioneering operations in Malaysia, using a fixed-wing aircraft and a helicopter in combination, revealed a rich potential.

Existing clouds are not required. These techniques generate their own clouds, from zero if need be. New technical resources to deal with droughts exist in the here-now, together with the unique operating acumen essential to effectiveness and efficiency.

WORLD HISTORICAL RENOWN AWAITS THE VISIONARY LEADER WHO FIRST USES THIS TECHNOLOGY, AGAINST DROUGHT PROBLEMS THAT HAVE DEFEATED AND HUMILIATED CONVENTIONAL "HIGH TECH."

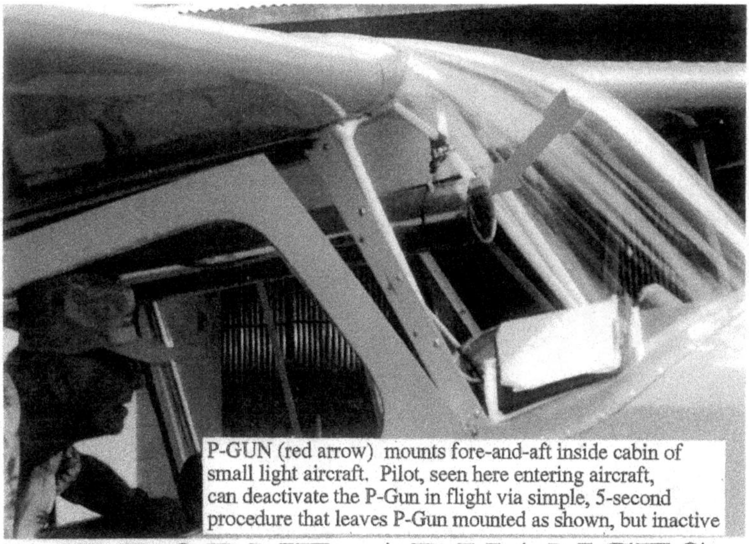

P-GUN (red arrow) mounts fore-and-aft inside cabin of small light aircraft. Pilot, seen here entering aircraft, can deactivate the P-Gun in flight via simple, 5-second procedure that leaves P-Gun mounted as shown, but inactive

AIRBORNE ADVANCES

Throughout 1997-98, practical research continued with etheric rain engineering in the airborne mode. Successful airborne tests in Malaysia in October of 1997 proved that techniques pioneered in Hawaii were even more effective in equatoria. Rapid engineering of rainy weather, sometimes violent, made in-flight deactivation of the quarter-wave Bull units absolutely essential. These translators were originally designed for attachment to wing struts. Successful test flights in Malaysia and Hawaii proved the feasibility of moving the translators INSIDE the aircraft. Simultaneously, an exceptionally powerful quarter-wave Bull translator was developed and tested -- the fabulous "P-GUN." When mounted INSIDE a light aircraft, the etheric P-Gun functioned flawlessly, as though no aircrcraft were present. A rain engineering pilot flying solo can now deactivate the P-Gun in the event of heavy weather, enhancing safety.

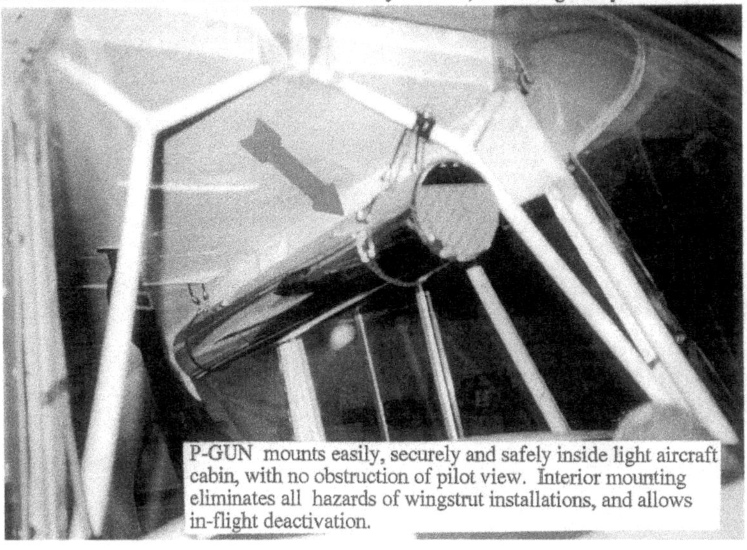

P-GUN mounts easily, securely and safely inside light aircraft cabin, with no obstruction of pilot view. Interior mounting eliminates all hazards of wingstrut installations, and allows in-flight deactivation.

P-GUN ON SKYVAN. Less than 4 ft. long, P-gun is dwarfed by huge cargo bay of Skyvan aircraft in Malaysia. Operations proved Skyvan too big and too fast for extended operations, even though flight with P-gun deluged Kuala Lumpur to end regional drought. April 1998.

HARD LABOR. Captain Bala Ratnam refuels his Cessna Skyhawk the old-fashioned way: from jerry cans. George Wuu studies Bull translator on aircraft's wing strut. Segamat Country Club in central Malaysia. October 1997.

A.E.R.E.O.

Airborne Etheric Rain Engineering Operations

RAIN GUN ALOFT. Wing strut rain guns like this, in Malaysia in 1997, using no chemicals or electric power, produced violent weather rapidly from fair conditions. Guns had to be modified for cabin installation, permitting deactivation in flight. Captain Ratnam was fully convinced of their power after several wild encounters with lightning strikes.

Helo with skid-mounted P-Gun at Langat Dam, Kuala Lumpur, Malaysia. April 1998.

HELICOPTERS

EFFICIENCY, FLEXIBILITY, RESULTS

A helicopter fitted with a P-gun is the ultimate etheric rain engineering modality. This was proven in Malaysia in April/May 1998, and in China in 1999. Nothing in more than 30 years of etheric weather engineering development can compare with a P-gun helicopter for RESULTS. Helicopter speed can be adjusted from the hover to more than 100 knots. Experience proved that optimum results in etheric rain engineering occur at approximately 50-60 knots. Helicopters are unmatched observation platforms -- a vital asset. Their ability to land virtually anywhere is protection against violent weather -- a frequent consequence of airborne etheric rain engineering. Combined operation of a helo and a light aircraft, both equipped with P-guns, allows thousands of droughted equatorial acres to be accessed and rained -- like crop-dusting with water.

P-Gun being mounted on helo skid, Langat Dam, Kuala Lumpur, Malaysia. April 1998.

2008 HONOLULU, HAWAII

Progress continues, as the arrow indicates the presence of the potent, 18-inch long MINIGUN . YES, INDEED, IT WORKS!!!!

HISTORIC FIRST. Colonel W.A."Willy" Schauer, USAF Ret., the first pilot in the history of the world to fly airborne etheric rain engineering missions. A former USAF fighter pilot with 31 years service, Colonel Schauer is described by TJC as "possessed of phenomenal powers of observation." Schauer flew all the pioneer missions of Operation Red Baron in Hawaii that established the feasibility and power of airborne etheric rain engineering.

TRIUMPH IN CHINA. TJC (r.) joins jubilant pilot Edmund Wuu of Singapore as the latter dismounts from helicopter rain operation on Hainan Island, China, in September 1999. Single helo pass with P-gun aboard, produced extensive rain on Hainan, which forced cancellation of follow-up flight the next day. It was still raining at take-off time!

FORCED DOWN by VIOLENT WEATHER. Thunderous conditions loom above research Cessna forced down to safety on Malaysia's Senai Airport. With ERE boss George Wuu aboard, Cessna had taken off shortly before in fair weather, with no severe conditions forecast. Gun-like translator seen on wing strut caused rain, thunder and lightning "Out of nowhere." Immediate return to airport was forced. Similar reactions came on several other test flights.

PIONEERING IN CHINA, SEPT 1999. First-ever airborne etheric rain engineering operation in the People's Republic of China, was set up by ERE and managed by technical staff of Hainan Airlines and its CEO, Mr. Chen Feng. Success was achieved with a single helicopter pass of 20 mins. Here, TJC waits on Sanya Airbase of the Chinese Navy.

OPERATION RED BARON, Hawaii '94. Typical etheric translators attached to the starboard wing strut of research aircraft. Red Baron program was a world first in applying etheric rain engineering from an aircraft. Combining geometry, velocity and direction, revolutionary modality uses no chemicals or electric power to engineer rain.

Colonel W.A.Schauer USAF (Ret.) boards aircraft in Honolulu, Hawaii (below) during tests of micro device (Arrow) for airborne etheric rain engineering. Tests were successful. ABOVE: Engineered RAIN INCOMING behind Diamond Head.

RAIN ENGINEERING HELICOPTERS AND AIRCRAFT

TJC demonstrated in Malaysia that a helicopter carrying etheric projector tubes as shown in the sketch, can be readily aligned with tropical ether flows, damming up the flows by flying straight into them. High potential etheric force, when engineered in this fashion, draws immense volumes of atmospheric water vapor to itself. Extensive clouds and rain ensue. Rains often last for many hours when triggered in this fashion. The extreme mobility of the airborne mode, permits targeting of droughted areas, previously one of the most difficult rain engineering tasks from fixed bases.

Improved design and efficiency of the simple translator tubes now suffice for a single "P-gun" to engineer rain. No chemicals and no electric power of any kind are required

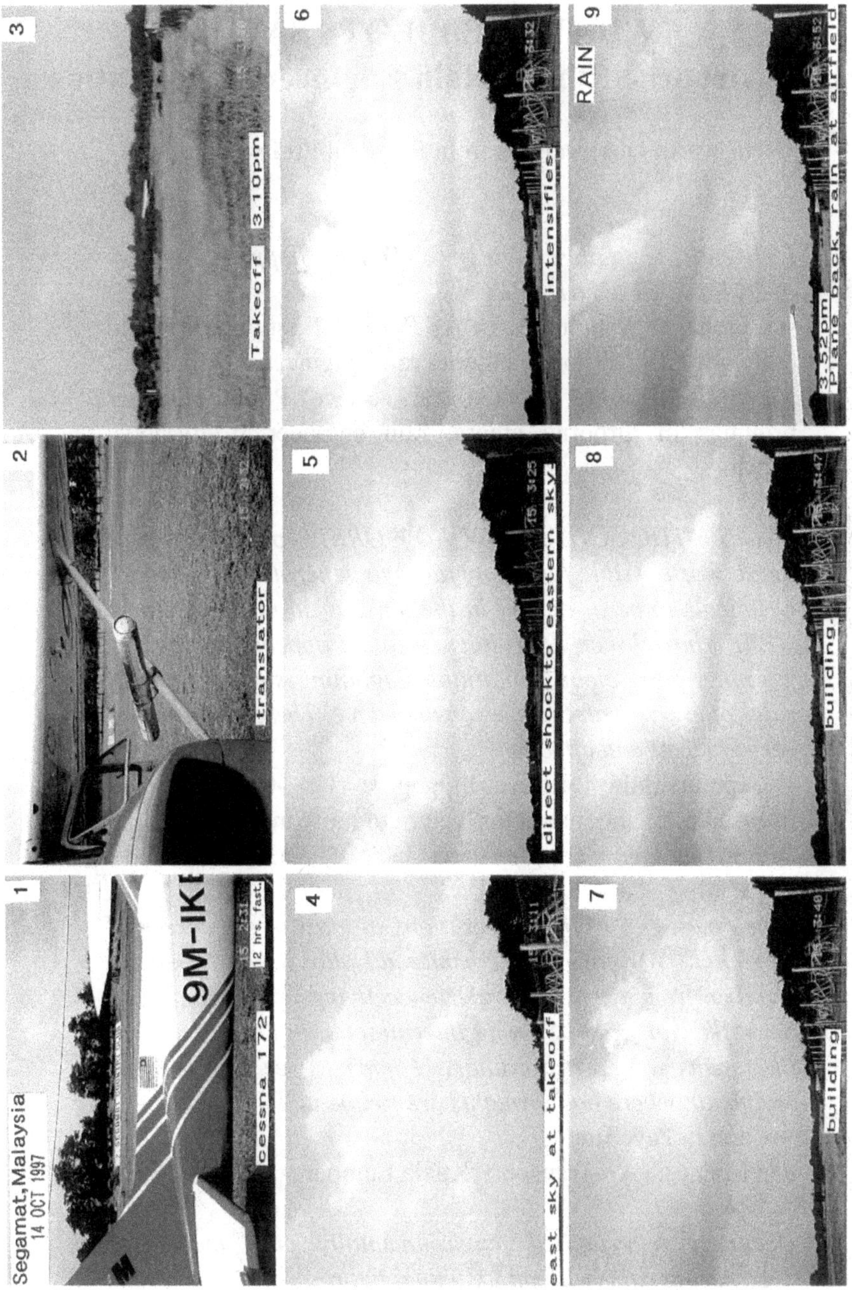

Rain engineered 42 mins after takeoff persisted thru 2am next day. Approximately 2 inches of rain over 250 sq. mile area.

WHAT THE PILOTS SAY
About Airborne Etheric Rain Engineering Operations

Real-world Observations from Practical Professional Men

I haven't the faintest idea of WHY rain happens in these operations. BUT IT HAPPENS!
—Colonel W.A. Schauer, USAF Retired. First pilot in world history to fly airborne etheric rain engineering missions. Vetern USAF fighter pilot (31 years service). Pilot for pioneer provings of airborne etheric rain engineering missions, Hawaii.

I NEVER THOUGHT IT WOULD WORK! I could not believe that it was possible for empty tubes to generate rain, with no chemicals to be dispersed in the air. But, in my experience, NEVER have I seen such violent weather appear, without any warning, from clear conditions. Lightning strikes on both sides of my aircraft further convinced me. These methods are very efficient and effective.
—Captain Bala Ratnam, Singapore. U.S.-trained, FAA-licensed pilot and instructor. Vetern of more than 20 airborne etheric rain engineering missions in S.E. Asia.

I have flown in tropical conditions in most of my career of more than 30 years as a commercial pilot. I know tropical weather from extensive experience flying commercially in Short Skyvan, out of Subang International in Kuala Lumpur. the effects were extraordinary. I hope to see this used in Myanmar, where bad droughts are frequent.
—Captain Paw Tun
Pan Malaysia Air Transport, Kuala Lumpur

Definitely a practical, environmentally pure means of engineering rain, with an unlimited future.
—Edmund Wuu, Singapore. FAA-licensed fixed-wing and helicopter pilot. Participant in etheric rain engineering missions in Malaysia and China, including history's first coordinated fixed-wing helicopter operations.

Chapter 7

HURRICANE AND TYPHOON CONTROL
The Anatomy of a Challenge

As a string of Atlantic hurricanes devastated Florida and many other areas of the eastern USA, numerous persons familiar with my etheric weather engineering work have asked what might be done to lessen these chains of calamitous hurricane events by using etheric technology. Conventional technology has no rational answer to the problem. Lunatic suggestions to explode atomic bombs against the eye wall of these immense vortexial systems attest to orthodoxy's dangerous bankruptcy, and also to orthodoxy's perpetual readiness to embrace the murderous.

Establishment thought and technology has no effective and safe palliative for hurricanes, and nothing in sight except H-bombs. Why? Orthodox thinking and the habits it produces function an octave too low – neurotically locked down in the "safe" zone of the lifeless. Hurricanes are life energy in high action. No access to life energy principles is possible for minds dragooned from puberty to accept the fraudulent teaching that there is no life energy and no ether.

As a practical matter, there is scant possibility that the necessary funding and resources would be made available to expand the successful work with hurricanes and kindred systems that I have done in the past, because of this fraudulent teaching and its consequences. Political leaders and their satellite scientists, as well as the media clones of our negative world, are structurally incapable of accepting or dealing with the life energy or its technological use. That is the main reason why airborne etheric rain engineering and its demonstrable capabilities have been evaded for more than a decade in a droughted world. Countless reels of time-lapse videotape, showing the life energy being accessed technically and used technologically to engineer rain, evoke no rational scientific interest – only the desire to negate, evade and avoid the physical reality of life force.

Reactions of this negative kind pervade the scientific world wherever the life energy is revealed by any kind of technical activity, such as diverting or deactivating hurricanes, that brings the life energy into use, focus or consideration. These negative attitudes and reactions – the operative effects of mankind's most serious structural bias – have retarded human progress by centuries. We earthlings are centuries astern of where we innately belong and could be, were it not for the predominant negativity of the earth plane and the hidden powers sustaining that negativity.

Dealing with hurricanes gives us a contemporary opportunity to illuminate how this kind of bias – now virtually an affliction of our species – retards progress and condemns millions of humans to unnecessary losses, suffering and misery. The core of the problem is not so much technological as it is emotional and neurotic. The real-world violence of hurricanes is an ideal backdrop against which to examine the pathological formatting that lavishly sustains imperial foreign wars, while recoiling from the massive manifestations of vital energy behind hurricanes. The warfare is rationalized. Only the hurricane is terrible.

The particular etheric force that creates the hurricane's eye goes also to the roots of our own lives. The motional physics of the typical hurricane are a macro-replication of the bodily human superimpositions that produced everyone reading this, starting with the "eye" of their conception! The blocked organismic movement of this etheric force – the life energy – is the source of personal behavioral problems in most people, as the late Dr Wilhelm Reich elucidated for mankind before his death in an American penitentiary, and the burning of his scientific books and papers by the American government. The Children of Tomorrow must never know what makes their hearts beat. The same force might propel a car, or light a house, or replace wired electricity and gasoline.

Dr. Reich demonstrated clinically that when the organismic life force moves strongly in a body accustomed to life energy stasis, terror is experienced. This is a consuming terror that the afflicted will do anything to suppress – including bolting from therapy. While this is an individual scenario, its essentials are conveyed to the organizations and enterprises that such blocked individuals ensoul. In a typically perverse fashion, therefore, great institutions and organizations would rather have the hurricane than have it modified by life energy laws and technology.

Diverting, diminishing or controlling hurricanes by methods rooted in the same etheric laws that create the hurricanes is thus resisted technically, behaviorally and conceptually. To show or demonstrate that the life energy can be used technologically, evokes irrational denials of objective, irrefutable evidence.

Confronted with unassailable time-lapse videotape, men who have been educated through a distinguished university will often respond with, “I don’t believe that.” Belief is readily enthroned over demonstrated fact by doctors of philosophy – when the life energy is involved. This reality has been encountered by me, and by the Singapore ERE organization, for nearly 40 years. That is why I do not chase after governments, offering programs to diminish or divert hurricanes, and why governments will certainly not chase after me.

The responsible institutions actually prefer dealing with damage, suffering and misery, than to dealing with the life force. That which is life-negative is neurotically comfortable. The life force to which I have devoted most of my earthly sojourn is uncomfortable, disquieting and disturbing to blocked personages. Great indeed is the fear of the living in our anti-civilization – itself an appalling sink of negativity.

In the now out-of-print, comprehensive history of my weather work entitled *Loom of the Future*, two factual accounts appear of two different modes of dealing with hurricane systems. These modes are the fixed-base operation and the mobile operation. In the case of Hurricane Iniki, in Hawaii in 1992, a timely and effective diversionary influence was exerted on Iniki by the writer that prevented its projected impact on the city of Honolulu, leading to reports of grateful amazement in the *Honolulu Advertiser* the following day. Shoved by etheric engineering a little to the west from its estimated course, Iniki hit the island of Kauai instead, with terrible consequences. Similar impaction of the monstrous Iniki on populous, tightly integrated Honolulu city would have resulted in a $15 billion damage bill, and thousands of casualties – at least. The writer was able to bring about this diversion of a massive hurricane through the use of a single piece of etherically actuated equipment on his penthouse deck, at Punaluu in northeastern Oahu – not even an ideal site from which to produce the desired diversion. None of the hurricanes that battered Florida in 2004 was as fierce as Iniki, with its 937-millibar eye and 150 mph winds.

Counter-action against Iniki was from a fixed base, utilizing etheric engineering principles. The technical knowledge involved was developed in literally thousands of shipboard rain and weather engineering operations conducted during 300 Pacific Ocean crossings. The practical efficacy of these technological applications of etheric force is attested to in numerous time-lapse videotapes. In eight years of residence at Punaluu, Hawaii, the writer also carried out extensive investigations of regional etheric force from his large penthouse deck, nine floors above the ground. This experience was invaluable in the Iniki incident.

The years of mobile, high seas operations on shipboard, making constant use of radar, equipped the writer with a basic understanding of how weather systems could be made to behave, within limits, by understanding and applying the operative laws of etheric force. A collision avoidance system, using information from the ship's two radars, left no doubt that the laws of etheric force overrode conventional expectations of weather system movement. Higher laws prevail in etheric rain engineering operations.

In *Loom of the Future*, the writer also describes a mobile encounter with a hurricane between Los Angeles and Honolulu, aboard the Matson container ship SS Maui, with Commodore Kenneth Orcutt USNR in command. He requested my intervention. This particular hurricane was approaching Honolulu from the east, in such a way that its northeast quadrant arrived on the ship's stern the evening before the Maui's scheduled Honolulu arrival. The island of Oahu was on hurricane alert for that same day. Simple countermeasures were initiated on the Maui's flying bridge, utilizing water-powered etheric engineering equipment of almost incredible simplicity. Procedures learned in years of practical experiment with etheric force were applied to a pending hurricane disaster.

On the ship's 3-centimeter radar, clearly discernible bands of countering force could be seen moving astern, directly away from the ship and perpendicular to the hurricane winds on our stern. These bands were pulses of etheric force, made tangible to radar by the water they had directly absorbed from the atmosphere. They were "backwashing" directly into the extremely high potential etheric force that was impelling the hurricane and creating the hurricane's wind flows. This radar scenario objectified our engineering intention to diminish, or perhaps completely remove, *one arm of the hurricane* – the arm on our stern. Every hurricane system is existentially dependent on having *two* atmospheric arms, each of which is underlaid and driven by a stream of etheric force.

These two arms superimpose to create the hurricane's eye and systemic winds – those winds being entrained by the etheric flows involved. The hurricane is an implosive etheric vortex that impels the atmospheric circulation we see objectified and tracked by meteorologists on TV.

By dawn, the Maui was in fair weather. The US Navy at Pearl Harbor had sent out an aircraft at daylight to precisely fix the hurricane's eye, but the system menacing Honolulu had vanished, engineered out of existence overnight.

This mobile etheric engineering experience and the fixed-base Iniki operation give a good introduction to what has been done factually in the past. This knowledge is capable of implementation – in the here-now – to influencing hurricanes. In the case of the typical, annual Florida series of hurricanes, simple, inexpensive diversionary bases can be set up along the Florida coast for inconsequential funds – peanut money. Such little bases would require no chemicals or electric power, and could be arranged to prevent hurricanes coming ashore, especially if their functioning were coordinated with shipborne etheric engineering equipment, and with airborne units. Such operations would need to be developed in practical formats, but would be far simpler in cost and complexity than, say, sending John Wayne to Iraq.

The many reporting TV meteorologists that we see standing in the rain in oilskins and sou'wester hats, looking like old-time ads for Scott's Emulsion, typify our current attitudes to hurricanes. The reporters appear drenched with rain and buffeted by the winds in full view of the TV audience. They epitomize and exemplify the *idee fixe* that we just have to stand and *take what Mother Nature hands out*. Not so.

The successful use of etheric engineering equipment aboard aircraft is already a demonstrated practical reality. No chemicals or electric power are utilized, just geometry, direction and velocity, plus access to the etheric continuum, with all its liberating portent for mankind. Airborne etheric rain engineering technology adds another potentially powerful dimension to any hurricane control measures. Particularly is this true in learning to take one arm off a hurricane in an airborne adaptation of our hurricane adventure on the SS Maui. Preparations for such positive action should be initiated without delay, building up experience and extending my basic discoveries before the next hurricane season. Such experience can be secured by undertaking the diversion and dismantlement of ordinary low-pressure systems that are available year round. The hurricane is the ultimate version of the low pressure system, which springs from the same geometric roots The alternative to such a constructive, rational, inexpensive approach to hurricane diversion is what we have now: just standing there and taking what Mother Nature hands out – like the TV reporters do in oilskins and sou'wester hats.

The cardinal technical principle involved in any hurricane control measures is to use the colossal energies of etheric force in a hurricane *against the hurricane*. A science that thinks an octave too low, with no understanding of what vital energy actually is, cannot access and technologically utilize that force. The chief characteristic of vital energy is that it flows from low potential to high potential – the reverse of conventional, standard concepts of energy potentials.

Under such a technical regime, relatively small, geometric devices, when appropriately shaped, mounted and directed, can exert quite a stupendous influence on a hurricane, via their etheric emissions. Elegant, effective, and employing no chemicals or electromagnetic radiation, such an approach stands in contrast to the dark threats that nuclear bombs will be cast into the eyewall of a hurricane. Presumably, that can only be done with all the required geometric parameters governing atomic explosions, such as time and position, suitably satisfied. Perhaps that isn't too easy within a powerful etheric vortex, and bloody dangerous as well. Furthermore, given the antagonism between concentrated vital energy and atomic energy that Wilhelm Reich demonstrated, the bomb might not go off.

The necessary preparatory work for the kind of constructive hurricane diversionary work I have outlined will take some time, plus protected status and minor financial resources. In my opinion, with a year of such preparatory work, our entire posture vis-à-vis the Florida hurricane season, would be totally transformed from terror to confidence that we were finally doing something meaningful about hurricane diversion and control, instead of just standing in the rain.

Before being swamped by "irrational exuberance" over these prospects, a necessary remembrance is that the scientist who broke through to the primal secret of the low-to-high flow law of etheric force, Dr. Wilhelm Reich, died in a Federal prison in 1957. The American government then burned the scientist's books and experimental bulletins containing knowledge necessary to combat hurricanes and the destruction they inflict on our country and on the lives of our fellow human beings worldwide. This miserable book burning occurred in America less than 60 years ago, paralleling Berlin in 1934, and demonstrating how little we learned from WWII. The hidden powers behind that New York book burning are with us still. Their wish is to prevent the wondrous powers of vital energy from liberating the human beings that they regard, and treat, as cattle.

Chapter 8

LETTER TO RADIO HOST JEFF RENSE

Excerpt from a letter written by Trevor James Constable to Jeff Rense, the international Internet talk show host, on 30 April 2003, following a show appearance. This concerns practical etheric effects in the earth environment:

Right on commercial break during the show, you raised the question of why trailer parks appeared to be especially vulnerable to tornadic conditions. Unfortunately, we never got back to discussing this. The reason that trailer parks and mobile home parks seem so vulnerable, in my opinion, is that from an etheric point of view, they are rows of *negative etheric accumulators*. As such, they are explosions waiting to happen, and requiring only the presence of a high etheric potential vortex, i.e. a tornado.

The standard orgone (etheric) accumulator is a six-sided metal box, covered with a layer of insulating material, preferably organic in origin. Adding layers of metal and organic material will increase the etheric potential within the box core, up to the capacity limit of the box. This is the essential nature of Wilhelm Reich's original invention, which shows that the 2nd Law of Thermodynamics is not valid in certain arrangements of materials.

Now, if you *reverse* the layering process and proceed from a metal exterior through a layer of insulating material, e.g. wood, the etheric potential of the core space will be *reduced*, rather than increased. This is a negative etheric accumulator, although I have not come across anything in Dr. Reich's work showing that he ever dealt with this. His focus was on the original accumulative arrangement of materials, and its complete verification.

In our ERE rain engineering work, we have made abundant use of the negative accumulator, using the principle to create the shooting action of our P-guns and other devices. Difficult to demonstrate or develop in desert or semi-arid locales, the principle is demonstrated incontrovertibly in maritime and especially in tropical maritime locales. In the airborne mode, it is the key to truly immense effects, when the motions of the aircraft are lawful and the negative space made resonant.

Transfer this basic knowledge now to the typical trailer park in Tornado Alley. Mobile homes and trailers are mostly metallic structures. Many of them are of unpainted aluminum or dural. Directly underneath the metal skin is a thick layer of insulating material, usually glass wool backed with aluminum foil. This may constitute a second layering arrangement, depending on the particular manufacturing processes employed. The fibre board lining of the coach is attached over the aluminum backing of the glass wool. The entire coach thus becomes, essentially, a 2-fold negative etheric accumulator. The ambient etheric potential inside the coach is below that of the surrounding atmosphere. Great care is taken by the manufacturer to exclude drafts and make the structure as comfortable and as weatherproof as possible. These measures increase the efficiency of the "negative accumulator" that has been unwittingly constructed.

The difference in etheric potential between the outside atmosphere and the ambient etheric potential inside the coach is not great enough to pose any danger under everyday conditions. Normal venting actions with doors and windows support a continuous stabilizing commerce between the two etheric potential "zones." Not so in tornado conditions.

Mobile home park residents "batten down" for what is to come. This battening down must tend to lower to its minimum – the etheric potential within the coach. Battening down must also tend to limit any commerce between interior and exterior etheric potentials. The negative accumulator (the coach) now loses its stabilizing linkage to the outside atmosphere.

Enter the tornado into the format. *Outside*, there is now a super-active vortexial system of extremely high etheric potential. This high etheric potential is what causes power transformers to explode. The etheric potential difference between the interior of the coaches and the outside atmosphere increases rapidly. The lower potential inside the coaches becomes a chain of *feed points* for the extreme high potential of the tornado system. The tornado does not need to physically impact the coaches. Since etheric force flows from low potential to high, the low potential within the coaches expands with volcanic force as it seeks to feed the voracious high potential of the vortex outside. Etheric force moving under such extreme differentials will readily penetrate the metal skins of the coaches, or split them.

My opinion is that this is the reason why you see so many mobile homes and trailers literally explode near tornadoes. This is also the reason why tornadoes seem to "favor" such parks, where negative etheric accumulators are set in *geometric rows*, usually oriented to the cardinal directions, i.e. north-south and east-west. Surely, it is a tragic scenario when you see – time after time – rows of such homes blown asunder.

In due time, scientific understanding and elucidation of the existence and properties of the ether will lead to changed technical protocols for mobile home design and construction. My opinion is that their vulnerability to tornadoes can be designed out of such coaches.

On some future visit Jeff, we can perhaps present these enlightening ideas.

There are ancillary aspects to it also, in the kindred essential character of airliners aloft. They are flying negative accumulators, as study of the foregoing will clearly indicate. Lots of interesting possibilities now arise for discussion, including the rapidity with which stewardesses age and other intriguing spin-offs and consequences. All this would have long ago received the scrutiny of physics and especially biophysics, had the formal world not permitted Einstein to paralyze their thought processes.

Chapter 9

HIGH ETHERIC POTENTIAL
vs.
HIGH VOLTAGE ELECTRICITY

During the summer weeks of 2003 there were many instances of tornadic activity in U.S. areas. Clear video was recorded of tornadoes making their way across open country and into towns and villages. Etheric phenomena integumented in these scenarios stimulated many correspondences with my own career in manipulating etheric force for rain engineering and related purposes. Typical armored perception, in the media, did not register the frequent spectacular antagonism between high voltage electricity and high potential etheric force in many of these incidents. Their essential character eluded commentators watching these happenings, as commercial transformers exploded before their eyes.

The most vivid phenomenon recorded showed a sharply defined, spinning tornado funnel advancing across the countryside, with power transformers literally exploding well ahead of its line of advance. The tornado did not need to physically contact the transformer in order to cause this explosion into a white flash of flame resembling a huge photographic flashgun. In another instance recorded on video, the tornado is advancing toward the video camera and while it remains perhaps 100-150 yards away, a power transformer in the left foreground explodes in a white flash, as previously described. There is no visible arc or bolt from tornado to transformer. Elsewhere at other times I have seen other video recordings of these transformer explosions in the near-presence of tornadoes. I have grasped what was basically involved for decades. Only a few humans, aware of the ether's physical existence, understood that the ether was "talking" to us in its own language – the mode of its manifestation.

Since the time of the late Dr. Wilhelm Reich, a definite antagonism between electrical activity and high etheric potential has been acknowledged and noted in the history of *orgone energy* (Reich' s name for ether). A compelling example occurred in the well-known Oranur Experiment at Reich's laboratory in Maine. Radioactive cobalt there, which can be deemed extreme electrical activity, was brought into contact with the high etheric potential of Dr. Reich's specially designed and constructed orgone room. The latter was a large, metal-lined, room-sized orgone accumulator. Orgone energy concentrations within this room were considerably higher than in the surrounding outside atmosphere. Drastic and complex reactions ensued. One medical doctor, Reich's own daughter, nearly died from exposure to these reactions. Further objective consequences included the buildings of Reich's research facility glowing at night, testimony to which was borne by the late Dr. Elsworth Baker, a distinguished psychiatrist and author, who became head of the College of Orgonomy after Reich's death. Further details can be found in literature about Reich's life and career. These details need not concern us here, other than citing the Oranur Experiment as a starting point where antagonism between etheric force and radioactivity was first noted by professional people.

Only when etheric potentials become high – much higher than is normally present at the level where our planetary life unfolds its activity – does the ether's antagonism toward high electric potentials become evident. This antagonism occurs when the extremely high etheric potential of a tornado vortex clashes with the high electric potential within a commercial power transformer. The electrical activity is then explosively suppressed, accompanied by a massive flash of light. This happens without the tornado necessarily touching the transformer, but rather, at some distance as occurred in the aforementioned videos recorded in the 2003 summer.

Only rarely does etheric potential reach levels sufficient to suppress electric high potentials. I have seen it happen in a somewhat unique physical situation aboard a merchant ship in a typhoon. Etheric potentials in a typhoon become drastically elevated as the barometer goes down. (Etheric potential and barometric pressure have an inverse relationship). Within a typical marine radar, the high electric potentials necessary to key the radar's magnetron are produced by a pulse transformer. Typical marine pulse transformers are approximately a cubic foot in size – a metal box containing the step-up windings that produce the necessary 30,000 volts or more. This metal box is filled with phenolic, which is essentially a solid block of phenolic and encases and insulates the transformer windings. The high voltage is passed to the remainder of the radar via heavy, porcelain-encased terminals on top of the box. Every technical precaution is thus taken to ensure that the 30,000 volts does not arc out of the metal transformer case.

What about something *breaking into* that phenolic-filled box?

That is what can happen when the entire ship, radar and all, becomes surrounded with high potential etheric force – as in a typhoon. High potential etheric force goes through material substance as though it does not exist, including the heaviest insulating phenolic. In one instance in which I was personally involved as a marine radio electronics officer, our radar's pulse transformer failed, instantly rendering the radar inoperative. There was no lightning strike on the ship or in the wheelhouse or on the radar indicator chassis when it happened – just a sudden failure of the radar.

When the pulse transformer was removed, a strange phenomenon was found. There was a hole right through the metal casing of the transformer on one side. This hole was *punched in*. An irregular, mole-like burrow, a quarter-inch in diameter, led from that hole right through the solid phenolic to the output terminal of the transformer. The punched-in hole and burrow evidenced a charge or force from *outside*, going right through the metal and the phenolic, and right to the point of highest electric potential within the transformer. The ship's engineers later obliged me by cutting open the defunct transformer, revealing the detail I have described.

Numerous instances of etheric force immobilizing electrical circuitry exist in the phenomenology of UFOs, which are etheric in origin. Aware of such instances, and also of the Oranur Experiment, the shipboard experience with the pulse transformer burnout was a direct personal lesson in this kind of elemental antagonism. My opinion is that phenomena in this category of antagonism between the two energy forms contains a definite living element. When the etheric potential becomes high enough, it "goes for" the highest electric potential as though determined to extinguish the electrical activity. Even a five-watt mentality can see how useless electronic and electrically-based weaponry would be, in an environment dominated by etheric technology – as portended by certain UFO phenomena.

In conversations with other veteran radio officers later, I found that several had "lost" their radars in extreme weather conditions via pulse transformer burnout. Unless exposed to the clearly penetrant power of extremely high potential etheric force, a pulse transformer could normally be expected to last the life of the radar in which it is installed.

Earlier, I outlined the etheric force factors at work in the special way that mobile homes and trailers in parks simply seem to explode in the proximity of tornadic activity. The tornado does not have to descend directly on the mobile home. Etheric potential within the home is lowered below the etheric potential of the surrounding atmosphere by virtue of its metallic exterior sheath and interior insulating surface. The typical mobile home is thus a negative etheric accumulator following the law of etheric potential, which is that low potential ether flows to higher potential ether – exactly opposite to electricity. Violent expansion takes place from within the mobile home to the extremely high etheric potential of the tornado. The mobile home therefore explodes remotely from the actual vortex. Mobile homes in parks are often seen to blow apart at the approach of a tornado.

A kindred phenomenon appears as the extremely high potential of a tornado explodes commercial power transformers in the tornado's vicinity. Etheric force penetrates all physical substance, including transformer casings and mobile home shells, which the ether can readily penetrate. These are not isolated happenings. Sufficient video recordings now exist of this phenomenon for scientific studies to be made. Scientists will find that explanations that exclude the ether and its basic laws from such happenings are fundamentally flawed.

Related effects occur with airborne etheric rain engineering operations. When a typical "P-gun" is used on a light aeroplane or helicopter in a rain engineeering operation, a major etheric flow is interdicted by correct use of this ultra-simple device. Local blockage of a vast etheric flow that actually goes right around this planet takes place. When that immense etheric flow is locally and temporarily blocked, etheric potentials within that blocked area build up rapidly, because even slight elevation of these potentials is immediately reinforced from beyond the blockage point by a further instreaming from the main flow. This rapid etheric potential build-up condenses water from the atmosphere in prodigious amounts – visible as imminent rain. Such build-ups have been abundantly recorded on time-lapse video. When the wise and experienced airborne etheric rain engineer sees the build-up commencing, he will shut down his translator, turn his aircraft around, and get back to base with all speed.

This kind of engineered rain seems to come from out of nowhere, and has no connection to any formal meteorological theories or predictions. All our professional pilots in ERE Singapore have experienced such rains, rapidly materializing contra-forecast and contra-expectations. Thunder and lightning often accompany such operations. The experience can be chastening, especially for anyone hewing to the obsolete view that there is no ether.

Once a substantial buildup of etheric potential like this is produced by airborne etheric rain engineering, the effects of the engineered blockage of a major flow may recede "backwards" from the blockage site for perhaps a hundred miles. When this blocked flow recovers, and resumes, extensive rainfall is likely to accompany resumption of its normal passage, along with turbulence and hours of rainfall. Sometimes there are consequences for electric power systems over the areas where the original engineering was carried out. Some of these consequences can include happenings eerily similar to the explosion of the shipboard pulse transformer described earlier. The "aroused" ether goes for the highest voltage transformers, and other apparatus, and seeks to extinguish their activity.

An investigation of any sources of high etheric potential should always attend a regional power blackout, such as occurred in the USA in the late summer of 2003. Up until now, formal science has hidden behind some thoroughly disreputable shibboleths, such as "there is no ether." The antagonism between high potential etheric force and high voltage electricity is starkly real. Scores of thousands of volts of electricity can be immobilized and extinguished by high potential etheric force. There are scores of encounters with UFOs, dating from the earliest modern days of the phenomena, where the ability of etheric technology to suspend electrical activity has been demonstrated. This is a scientific arena that calls for the entry of dauntless men and women of vision, who will fight for the earth's etheric tomorrow. With AEREO, Airborne Etheric Rain Engineering Operations, we only have touched the fringe of this revolutionary technology, localized and primitive. Crude or not, without the use of chemicals or electric power, the ability of etheric force to condense thousands of tons of water out of the atmosphere is a pregnant technical happening.

The Civilization of the Universe undoubtedly has the technology in sophisticated advancement. That much may be fairly inferred from ultraviolet-sensitive videotapes shot by NASA 300 miles above the earth, and now in the public domain. Drastic and extensive revisions of the mechanistic world-order are in the immediate human future. Obsolescent irresponsibles in big jobs can be counted on to resist this inevitability.

Chapter 10

DR. RUTH B. DROWN'S RADIOVISION
Update and Extension

Researchers in the 21st century attempting to replicate Drown Radiovision usually attempt to work with the published patent for this device. Nobody has yet reproduced a functional Radiovision unit from this particular base. As one who was closely associated with Dr. Drown in her last days, my recommendation is that future researchers simply discard the published Radiovision schematic, including the photo-electric cell component, not only as obsolete, but an encumbrance to clear understanding of Radiovision as Drown actually carried out the process herself, immediately before her arrest, persecution and consequent death. As a personal witness to these events, right in the darkroom with Dr. Drown, my account can be considered authentic. My own long practical involvement with etheric technology, beginning in photography, helps further clarify the Radiovision process as RBD left in the world.

Primary in its influence on the thinking that must accompany a fruitful exploration of Radiovision, is the recognition that Drown Radiovision photographs are depictions of the *etheric body*. These photographs are not of the physical organs, glands and tissues. The etheric body is a subtle, unseen duplicate of the physical body in all its parts. The etheric body *animates* the physical body, and its energies are indispensable to the living existence of the physical body. Without the ether body, a corpse appears. Death is the withdrawal of the ether body.

The energies of the ether body are accessible to the seemingly simple instruments and medical technology developed by Dr. Ruth B. Drown. Each organ, gland, tissue and fluid of the human body has its own distinctive etheric rate of vibration, always the same, until the onset of disease. The Drown rates, thousands of them, were empirically established over decades of her work, for the entire body as well as for most diseases and pathological conditions. While these Drown rates correlate numerically to the Atomic Numbers on the accepted Periodic Table, they also connect the human body to complex functions of the Cosmos – as they are theoretically postulated and expounded in the esoteric Qabala. The human body is thus integrated with the Cosmos of which it is both part and product. Ruth Drown elucidated all this for suffering mankind – leaving humanity a monumental legacy.

The entire Drown technology is *solidly anchored to mathematics, and is one hundred percent a numerically-based approach to diagnosis*. This entire breakthrough will take wings when men and women find the courage and inspiration to synthesize these discoveries with the computer age and its burgeoning technology. Ruth Drown laid it all out for them, and now, they have to do a little work themselves, and take a few hits from an Establishment that is crumbling before our eyes.

These Drown numbers are etheric numbers, and the investigator of today must begin by recognizing that without the etheric presence, without the animation that the etheric world imparts to the physical world, we have only dust. From the Qabala we learn a technical fact vital to understanding Radiovision. As various densities or Sephiroth, descend from the Great Unmanifest to the physical plane, each Sephirah is *positive in polarity to the sephirah following*.

Applying this technical fact to the relationship between the etheric plane, and the following physical plane, the etheric plane is *positive* to the physical, and *positive to physical light*. This is why the etheric body or etheric sheath cannot be directly perceived by normal human sight, unless some process removes our optical limitation or otherwise brings the etheric events to our perception. The etheric body can be irradiated electrically in various ways to make it perceptible, the well-known Soviet work done with what they called "bioplasma" being an example.

Let us now transfer this knowledge to the final reaches of Radiovision as practiced by Dr. Drown:

On making a Drown diagnosis, she would at its completion have identified the *first cause* of the patient's problems. Symptoms tend to cascade through the organism, manifesting in various misleading symptoms that are secondary to the First Cause. To keep this illustration anchored in familiar territory, let us consider the well known Radiovision photograph of an abscessed molar tooth in a patient. The tooth rate and the rate of the attendant abscess are set

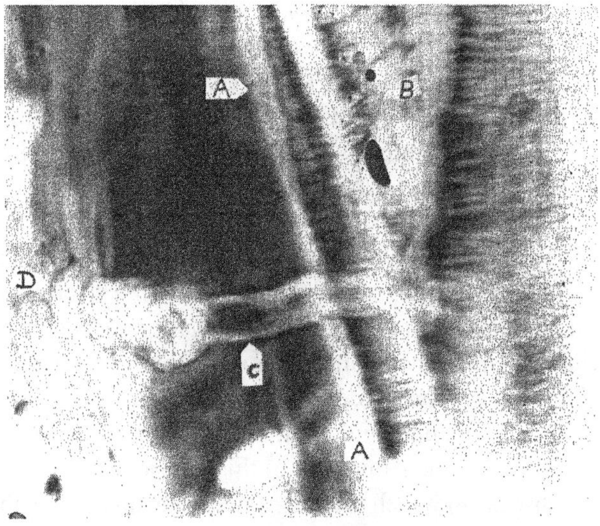

Drown's Radiovision photo of an abcessed tooth molar, made by tuning in with Radiovision instrument on the abscess. *The line of tooth structure is shown by the two arrows A, with the tooth root pressing into the gums. The photo is* cross-sectional, *and arrow B shows veins in the central pulp of the tooth, with black areas where blood flow has been sectioned in the veins by the Radiovision tuning. Arrow C shows a sinus of infection from the lower part of the pulp through the gum of the patient. Arrow D shows the gum boil as a bulge in the patient's mouth.*

up on the diagnostic instrument, and may be considered present at the instrument output. The entire assembly is now "tuned" to the problem. Under the original regimen, this complex "signal" would be used in the photographic attachment. Under the updated Drown approach, she found it was possible to *dispense entirely with the photographic attachment.*

Dr. Drown found that is was necessary only to clip the diagnostic instrument output briefly to one corner of the standard 8 x 10 Agfa cut film that she used for Radiovision. When this was done, the image-bearing output of the diagnostic instrument would be impressed in the film emulsion. That this image-impression may have been due to the supernal gifts and talents of this incredible woman is one of the things that the researchers of tomorrow will have to establish. Either that, or they will find that such image impression is yet another of the bewildering array of capabilities natively belonging to the etheric plane. Most likely, the latter is the truth of the matter.

I personally witnessed Dr. Drown carrying out the process further, in her own darkroom, as follows: The simple process of momentarily clipping the diagnostic instrument output to the film's corner is carried out in total darkness. The "exposed" film is then conveyed to a tall, glass-walled developing tank of standard commercial type. The film is placed in a standard developer. After a period of about 40 – 60 seconds, Dr. Drown would turn on a naked, 60-watt incandescent bulb in a holder, and allow it to shine against the glass wall of the developing tank. This is a step that would, under normal photographic technique, result in a *non-photograph* – in complete obliteration of any image on the film. Dr. Drown would agitate the film in the developer until *images began to emerge out of the film emulsion.* These forms could be clearly seen in outline. At this point, Dr. Drown extinguished the incandescent lamp, and the film development was completed in the orthodox way, via stop bath and thence into hypo for fixing.

When the film was washed and dried in the conventional way, it could be placed on a standard, illuminated viewer and studied just like an x-ray, except that the Radiovision photo was *cross-sectional through both hard and soft tissue*. In the example chosen from hundreds of real-world Radiovision pictures, the abscessed tooth is shown as though vertically sectioned right through bone, pulp and gum, blood vessels and all, with abscess in the picture center. You look at one wall of this incision. This sectioning goes right through the abscess to which the instrument was tuned, the abscess showing a sinus of infection from the root of the tooth, through the gum, to discharge inside the patient's mouth.

The crucial key to these achievements is that the film was exposed to light while in the developer undergoing development. This was a complete *reversal* of standard practice. This caused the physical objectification of otherwise concealed and latent etheric images. This is one of the master keys to the revivification and renewal of Radiovision technology. Another vital point arises here: Dr. Drown greatly *simplified* Radiovision in her final days by getting rid of the photographic attachment in the well-known patent drawing.

This simplification was based on one of Ruth Drowns' favorite sayings, drawn from her profound knowledge of the etheric world and its functioning: "*Everything is here, now. All we have to do is tune in on it.*" Thus it was that with the diagnostic instrument tuned to the organ and the pathology, at the etheric level, the entire imagery was present at the instrument output. The inference is that once the etheric energy entered into some kind of chemical union with the emulsion, that composite *repelled* future interaction. The image of the etheric lay latent until activated by incandescent light while developing was in process.

The many years this writer spent in photographing UFO's – in full recognition that these are phenomena rooted in the etheric – put him in instant recognition of the polarity problem involved that Dr. Drown had solved. This writer's appreciation of Ruth Drowns' genius was further enhanced by this farewell demonstration of its elegant simplicity. Her final aspiration was to unite her work with the then incoming wonders of videotape recording. Motion pictures of functioning organs and pathologies were within easy reach for the prodigious genius, whose motivation was the service of humanity.

Among the potential uses for Radiovision first used during the final period before Dr. Drowns' arrest, was taking cross-sectional photographs of artifacts. This used the same principle as medical Radiovision, but without the complex prelude of medical diagnosis. As a practical example, the UFO subject has resulted in large numbers of fake and fraudulent photographs being published and circulated with the motive of confusing clear thought on this subject. If an authentic UFO photograph is placed on the input to a Drown diagnostic instrument, a "rate" or frequency can be established for that artifact. This is usually a long rate of many digits. A fraudulent or fake photograph will not have such a rate, and probably no rate at all.

Dr. Drown and myself checked many famous UFOs this way, and found that they produced no Drown rate at all, and had been faked. Others that had been panned and ridiculed by our corrupt media, proved to have long rates. Many of the UFO photos taken by Dr. Jim Woods and myself on the high desert had rates that identified them as biological organisms and not ships from other planets. In the case of one of George Adamski's famous photographs, showing a large, fusiform vehicle in space, this photo and the rest of his photos were panned by the media and other critics. We found these images to be authentic via Drown analysis.

The large, fusiform craft had a long rate. When a Radiovision photograph of this rate was taken by the means previously described, this craft appeared on the etheric photograph in cross-section. Compartments within the vehicle were sectioned and showed indeterminate objects within those compartments. Ramps leading to the outside of the vehicle were identifiable, including one ramp upon which a smaller or "scout" UFO was seen as a compact, glowing object, either in the process of launching or docking. Dr. Drown had this photo enlarged to 24 x 36 and showed it enthusiastically to friends. She also included it in a public display of Radiovision photos at a meeting she convoked at the Institute of Aeronautical Sciences in Los Angeles. I introduced this exhibit at her request. Most of these photos disappeared in the horrible turmoil of her arrest and consequent death. For more information on Ruth Drown and her amazing work, see my other book, *The Cosmic Pulse of Life*, 4th updated edition, 2008, and these web sites:

http://educate-yourself.org/tjc/ruthdrownuntoldstory.shtml
http://www.rense.com/general39/ruth.htm

The potential of this kind of remote photography in cross-section was without limit, and remains without limit to this day. The remote location of ships and submarines and missing persons, especially political persons, was carried out successfully in Drowns' own lifetime, but was concealed for justified fear of clandestine, official retribution. Future researchers please take note.

History will characterize as demented barbarism the cruelty with which men of earth treated this God-like lady in her eighth decade. Almost a half century later, formal science still pathologically evades etheric force. This despite their own major advances having been secretly retrieved from crashed extraterrestrial spacecraft that were propelled by etheric force, confounding our brightest minds. Dr. Ruth Drown boldly crossed this technical frontier decades ago, and gave humanity Radiovision. This technology *verifies itself* with inarguable cross-sectional images of organs, glands and pathologies. Were this earth plane not a sink of negativity, her gifts would long ago have blessed suffering mankind.

" Everything is here, now. All we have to do is tune in on it."

Chapter 11

THEY CALLED HIM "HOBO"

The Little-Known Story of Percy Hobart

Major-General Sir Percy Hobart, K.B.E., C.B., D.S.O., M.C., Colonel-Commander of the Royal Tank Regiment, in a formal postwar photograph.

Winston Churchill threw down the *Sunday Pictorial* on the morning of August 11, 1940, with an angry scowl on his face. "We Have Wasted Brains!" blazed the headline to a slashingly critical article by Britain's top military analyst, Captain B. H. Liddell Hart. Dominating the page was a photograph of a hawk-faced officer in the black beret of the Royal Tank Corps, former Major General Percy Hobart. He was Liddell Hart's classic example of Britain's "wasted brains."

Practical pioneer and developer of the now-dreaded *Blitzkrieg* technique and former commander of the world's first permanent tank brigade, Hobart's revolutionary innovations in armored warfare had won him international military fame – and special attention in Germany. Dire peril now threatened Britain, but General Hobart was not commanding British tanks. He wasn't even in the Army. He had been found serving as a corporal in the Home Guard [over-age men and other civilians otherwise unfit for regular military service, meagerly armed, whose "uniform" was an arm band] – the highest responsibility Britain's military mandarins were willing to give to the progenitor of the *Blitzkrieg*.

Aroused by Liddell Hart's exposure of the situation, Churchill was determined to change Hobart's assignment. In the process, the prime minister was to launch and bring to its climax a drama of personal resurrection unsurpassed in military history. As Churchill pressed buzzers and rumbled memoranda to his secretaries, the country stood on the brink of ruin. The struggle with the Luftwaffe raged overhead. German armies were massing on the French coast for the projected invasion. The British Army had been routed in France with the modern tank methods first demonstrated to the world by Hobart, now a Home Guard corporal. The Germans had learned and applied only too forcefully the techniques pioneered by Hobart's tank brigade years before.

The prime minister directed that Hobart should be taken back into the Army. The chief of the Imperial General Staff should give him at least one of the new armored divisions to command. Delay was to be avoided. A personal meeting was to be arranged promptly between Corporal Hobart and the prime minister.

In a modest home near Oxford, lean, bushy-browed Percy Hobart was preparing to leave for his Home Guard duties. The one-time general who had commanded hundreds of armored vehicles in maneuvers and raised and trained the 7th Armored Division in North Africa took a wry look outside his front door at what was now his "transport." A baby Austin driven by a member of the Women's Volunteer Services stood waiting. The telephone jangled, Corporal Hobart answered, and found himself talking to one of Churchill's secretaries. The tank expert was asked to have lunch with the prime minister at Chequers, the official country residence of the British leader. Bigger things were in store for the aggressive 55-year-old ex-general, whose stormy and controversial past held the key to his future.

From the early 1920's, when he had transferred to the Royal Tank Corps as a military engineer, Hobart had turned his thinking to the future. He was among the few pioneers in every major nation to whom the tank appeared as the decisive land weapon of any future war. These tank enthusiasts, British, German, American and French, took their tactical inspiration from two outstanding British theorists, J.F.C. Fuller and Captain B.H. Liddell Hart. Liddell Hart in particular was influential. He was even then winning recognition as Britain's leading military brain – in or out of uniform – and he wrote forcefully and persuasively in favor of the new doctrine of strategic mobility. This concept is basic to today's military teachings, but it was heresy in the 1920's. Liddell Hart held that tanks would restore to 20th century warfare the ancient Mongolian idea of extreme mobility – the Mongols' main instrument of conquest. Bloody slugging matches in the 1914-18 fashion were doomed. Generalship would again flourish and replace the dull butchery of mass frontal attacks by infantry.

Orthodox military minds of the time could not grasp such concepts, which demanded creative imagination no less than military understanding. Men with imagination, vision and ability to carry these qualities over into practical soldering were rare in the static-minded, socially centered British Army. Percy Hobart was one such man. His diversified background and interests ensured that imaginative, mobile thinking would be second nature to him. A student of history and its lessons, he had delved also into such creative non-military fields as painting, literature and church architecture. Vibrant facets of mind to which regular military life gave no scope, sparkled brilliantly in Percy Hobart.

Percy Hobart, right, in conversation with Gen. William Simpson, US 9th Army Commander. Units of Hobart's 79th Experimental Armoured Div. served with distinction with Simpson's 9th Army. The two men became personal friends. Simpson called Hobart "the most outstanding high British officer I met during the war."

Liddell Hart's "Mongolian" concept of strategic mobility became the focus of Hobart's considerable intellectual resources. Development of these concepts and their adjustment to the mechanical twentieth century dominated Hobart's life from the time they were put forward. His creative imagination had been fired by the military revolution he could visualize, but his creativity was combined with a rock-hard realism. "Wars cannot be fought with dream stuff," he used to say, as he poured his life's energies into the development of practical machines for armored warfare, and the effective methods of directing these new mobile weapons. His goal was to break military science out of the straitjacket of trench warfare by updating the Mongol methods.

Where the Mongols lived off the country through which they ranged, Hobart planned to carry sustaining rations in the tanks. Refueling would be from lightly protected dumps in the enemy rear, where the far-ranging armored columns would penetrate and strike. He worked with relentless zeal to cut "the tail" of non-fighting service vehicles which hobbled behind and almost immobilized conventional army units. Tank forces of the future were to be self-contained for the maximum possible range.

Down-to-earth problems such as these did not prevent Hobart from taking a prescient look up at the sky. He planned for the time when the increasing power and versatility of aircraft would permit mobile armored columns to be completely supplied by airdrop. Standard practice today, this concept was in those times often the subject of mockery. Hobart planned to send his hard-hitting columns ripping into enemy supply lines and nerve centers in the rear, paralyzing command and demoralizing troops in the front lines. Less than twenty years later, America's General George S. Patton was to carry out these tactics on a vast scale and with historic success.

Resistance to these radical ideas began to stiffen. The old order found its neurotic and professional security threatened by the progress of strategic mobility. "Hobo," as he was affectionately called by his intimates, viewed the old order and its resistance to the new ways with direct and unconcealed contempt. "Why piddle about making porridge with artillery," he said, "and then send men to drown themselves in it for a hundred yards of No Man's Land? Tanks mean advances of miles at it time, not yards!"

Views like these were shared only by a small military minority. The powerful ruling faction of military conservatives was convinced of the value of the tank only in scattered use to support infantry formations. Horsed cavalry had been literally swept from the battlefield by the machine gun, but cavalrymen and cavalry philosophy nevertheless still ruled the high commands of the British Army. Men like these regarded Hobart's ideas as anathema. Professionally, they were maintaining the kind of army that could fight the First World War over again. Content with familiar ideas and concepts, and fearful deep inside that Hobart and others might be right, these controlling conservative elements closed the high commands of the British Army to tank advocates.

During this same period in the USA, despite the nation's massive mechanical heritage, a similar situation prevailed. Development of an independent armored force was stifled on that side of the Atlantic, although General Douglas MacArthur held a vision of the military future similar to that of Percy Hobart. Tank development was largely left to devoted individual officers in both Britain and America.

What Hobart's faction lacked in authority was made up for with energy and persistence. Aided by the strong independent voice of Liddell Hart, the tank enthusiasts were finally able in 1927 to pressure the British military hierarchy into the formation of an "experimental mechanized force." Maneuvers demonstrated dramatically that such a force outclassed old-style formations, leaving them bewildered and embarrassed. The theories of Liddell Hart and Fuller and the practical genius of Hobart's training organization were vividly vindicated. The writing was on the wall for the old order.

The die-hards reacted with a more energetic campaign against tank advocates and theorists. At all costs tank men were to be kept out of high command. Major-General J.F.C. Fuller, whose writings had been widely acclaimed both in the US and Germany, was the first victim. By a series of subtle maneuvers he was quietly squeezed into retirement and never allowed to hold an important post. Other tank officers were sidetracked and discriminated against professionally.

Hobart was now a rising power in British military circles, and conservative machinations were directed against him. He miraculously survived these early efforts at strangulation of the new ideas, and held a series of commands in the Royal Tank Corps. He worked out a basic modern battle drill for tanks, and used all his considerable powers of persuasion to get radio-telephones for his armored fighting vehicles.

Like most things for which he struggled, radios are indispensable to the military of today. A tank in today's armies would hardly be considered battle-worthy without radio. But Hobart spent months requesting, cajoling, demanding it. When the precious radios were finally obtained, Hobo was as happy as a child on Christmas morning. "Control is as important as hitting power, armor or mobility," he said.

With the radios came a new dimension in tank tactics. The basic equipment for a modern tank force was now at hand and Hobart began building up the techniques of command and control that were to rock the world. He made a sharp departure from the army concepts of leadership then in vogue. He believed in men knowing what they were seeking to accomplish in a military operation, right down to privates. "I do not want automata serving under me," he told his subordinates.

He brought everyone serving with him into intimate contact with the higher strategic and tactical principles he was striving to establish in modern war. Although not an orator, Hobo was possessed of a virile and inspiring eloquence that generated tremendous enthusiasm. His gift was to focus this enthusiasm on practical military matters, charging the mundane with a rare magic. Hobart carried this principle over into the civilian circles where equipment was being manufactured for his tanks. When he finally got his radios, he sought out the young woman scientist who ground the crystals for these long-awaited sets. She was set up in the tank turret beside Hobart and he showed her how hundreds of fighting vehicles depended on the accuracy of her work.

After the young woman had gone away visibly impressed by what she had been shown, Hobart turned to his brigade major. "What a dammed boring, awful job that girl has, grinding those crystals – but now she knows where we'd be without her." [1]

The soaring enthusiasm generated by Hobart's methods reached its zenith in the 1st Tank Brigade, formed in 1934 as the world's first permanent tank unit on modern lines. By this time a brigadier despite his radical views on warfare, Hobart was given command of this historic unit. He quickly infused the brigade with a booming *esprit de corps* unrivaled in the British Army.

Medium battiolions of Britain's 1st Tank Brigade, under the command of Percy Hobart, in close-order drill during military exercises in 1934.

Under his control at long last was the kind of formation that could conclusively prove the case for strategic mobility. Hobart lost no time. In a series of brilliantly executed war games, he proved the feasibility of driving to the enemy's rear with fast-moving armored units and completely disrupting enemy organization. He carried the revolution even further.

Hobart proved that armored units could both travel and fight by night. This innovation forced a complete revision of strategic and tactical concepts, for it placed old-style military units more than ever at the mercy of armored fighting vehicles. He firmly established the fundamentals of cooperation between tanks and air power, central to all that is done on the modern battlefield. He drove the 1st Tank Brigade hard. He knew how much could be proved and needed to be proved and that he might not be granted the time by his superiors. Continuing antagonism toward tanks, tank advocates and the new concepts of armored warfare characterized the high command of the army, and Hobart was never sure that his next war game would not be his last.

These unsparing efforts by Percy Hobart gave birth to the basic technique of the *Blitzkrieg*, the new mode of mobile warfare that was to bring nation after nation tumbling down and force Britain to the brink of defeat. The British high command remained irrationally prejudiced against the military technique that Hobart was unfolding. With a curious kind of intellectual detachment, most British leaders did not believe that the devastating effects of Hobart-style armored units could be carried over into actual warfare. Purblind views such as these aroused Hobart's fiercest antagonism: "What in hell is the use of having war exercises," he would fume, "when every lesson they teach us is ignored?" [2]

Skepticism about armor was reinforced by a lingering love of the cavalry horse. The logical passage of this beloved beast into military limbo was delayed and obstructed by its devotees. These men became opposed to the tank on emotional, sentimental ground, and found in Hobart a hostile, aggressive opponent. Horsemen nevertheless carried far more weight than tank men in British military life. Cavalry experts not only ruled the army commands, but had long tentacles into the body politic. Their influence was such that as late as 1936 the then secretary of war, Alfred Duff Cooper, apologized to the cavalry in Parliament for mechanizing eight of its regiments.

Hobart's achievements were running a poor second to the cavalry horse in Britain, but elsewhere they were undergoing dynamic scrutiny. A strong-jawed German colonel named Heinz Guderian probed with Teutonic thoroughness and an enthusiast's zeal into the lessons of every Hobart trial and exercise. Every report, observation and paper pertaining to Hobart's force was meticulously analyzed by Guderian, the Hobart of the new German Army. These studies formed the basis of the new panzer divisions, armored spearheads of Germany's new army. Hobart's 1st Tank Brigade was Guderian's practical guide, and answered many of the German leaders' early problems. Guderian had his difficulties with German military conservatives, but he accorded his country's tank debunkers little attention. When they spoke of "tank limitations," Guderian would not listen. "That's the old school," Guderian would say, "and already it is old history. I put my faith in Hobart, the new man."

At the conclusion of some prewar maneuvers of Guderians' panzer division, the German general was reported to have offered a farewell toast in champagne – "To Hobart." The dynamic British pioneer was considerably less poplar in Britain than he was with the modern military men of Germany. Unreasoning conservatism was taking an even sharper stand against tank men than ever before. The irrational nature of the conservative

Heinz Guderian was Germany's most important architect of armored warfare. In the years before Hitler came to power, when tanks were forbidden to Germany under the punitive Versailles Treaty, he learned much about modern armored warfare from a close study of the pioneering work of Britain's military strategists. In his postwar memoir, he specifically acknowledged his great debt to the writings of J.F.C. Fuller and B.H. Liddell Hart. Guderian also carefully studied accounts of Percy Hobart's innovative tank operations. After Hitler's advent, these lessons were applied in rapid development of the world's most powerful and effective armored force. In 1934 Hitler sanctioned the new *Wehrmacht's* first tank battalion, and four years later he named Guderian to command Germany's armored formations.

General Heinz Guderian in his armored command vehicle during operations in France, June 1940. His panzer units played a major role in routing British and French forces in May and June 1940.

standpoint, combined with the menace to his country and disasters that he could already foresee, had turned Hobart into an explosively fierce advocate of what he knew to be true and proved by actual test.

Hobart, top left (in beret), Commander of Britain's 1st Tank Brigade, atop a modified 16-tonner during 1934 military exercises on southern England's Salisbury Plain.

The slender general's personal forcefulness and vehement manner of expressing himself in pursuit of his goals had earmarked him for professional extinction. "No man is any good who has no enemies," was one of Hobart's credos.[3] By the late 1930s he had more bitter foes in Britain's War Office than any other officer in the British Army. He had become involved in heated arguments with all of Britain's military mandarins. Every leader from the Chief of the Imperial General Staff downwards had felt the whiplash of his tongue and the weight of his eloquent logic. Confrontations with senior officers could not long continue. Hobart's passion for the armored idea was actually leading him to risk his all.

Efforts to tone him down had little success. A deeply concerned Liddell Hart, in company with General "Tim" Pile – another long-time tank advocate – took Hobo out to dinner one evening. Their purpose was to save not only Hobart himself, but the armored idea, which Hobo's confrontations with high personages was placing in jeopardy. Relaxing in a pleasant atmosphere, Liddell Hart quietly stressed to Hobo that he was alienating potential War Office converts by his infuriating ways of argument. Like all strong personalities, Hobart could pass from one extreme of behavior to another. Force was balanced in his character by a courtly and irresistible charm. "He apologized disarmingly," Liddell Hart recalls, "and promised that it would not occur again. But only a week later the Chief of the Imperial General Staff complained to me that Hobo had again been intolerably rude to him. I tackled Hobo about it, but he was completely unaware of having been rude to anyone."

In this climate of clash and controversy, Britain tardily began the formation of its first modern armored division. The Germans already had four and were building more. Hobart's fears and predictions were being realized. He was the logical man for the command, and the new secretary of war, energetic, reform-conscious Leslie Hore-Belisha, was determined that Hobart should get the vital assignment. War Office conservatives dug their toes in and treated Hore-Belisha to a bewildering exhibition of bureaucratic and professional resistance. The secretary was unable to put Hobart into the post, and recalled in later years: "In all my experience as a minister of the Crown, I never encountered such obstructionism as attended my wish to give the new armored division to Hobart."

A cavalryman, whose most recent assignment had been the training of riding instructors, was proposed by the War Office for command of the new armored division. This proposal fairly characterized the uncomprehending state of British military thought on the eve of the world's greatest war. In a compromise arrangement with the War Office, Hobart became director of military training. Hore-Belisha hoped by this stratagem that Hobart's personal drive, enthusiasm and knowledge of armored warfare could permeate all army training.

The tank genius was now deep in "enemy" territory. He was the last tank man of high rank left in an influential post. Like a loathsome infection, he was gradually walled off by the subtle processes of the War Office organism while pressure mounted to expel him entirely from that august body. Hore-Belisha was continually urged to dismiss Hobart.

The Munich crisis provided the right emotional climate and an excuse to get rid of him completely. He was bundled on a Cairo-bound aircraft, assigned to raise and train Britain's second modern armored division. With Hobo's removal to the Nile delta, tank thinking was exterminated in Whitehall [Britain's Foreign Office], and as Liddell Hart put it, "The British Army was again made safe for military conservatism." For these decisions on the part of its highest military professionals, Britain was to pay dearly in life and prestige.

Scattered motorized and mechanized troops with obsolescent equipment were all that Hobart found in Egypt as the basis for a modern armored division. A grim enough prospect in itself, the equipment situation was overhung by a demoralizing and obstructive emotional factor. Commanding in Egypt was one of the British Army's remaining conservative hangovers from the First World War, a soldier for whom Hobart, himself a decorated veteran of the first conflict, had never failed to express his professional contempt. The commanding general was also a socially-minded soldier. He especially detested Hobart at the personal level for his 1928 marriage, for which Hobart's wife had gone through the divorce court.

Modern minds would regard such a procedure as little more than a fact of life. To the British Army of the period between the wars, it was a transgression sufficient to bring many threats of professional retribution on Hobart, one of them from the general who now commanded in Egypt.

Hobart's arrival was followed by a brief and brutally unceremonious interview in the quarters of the commanding general. "I don't know why the hell you're here, Hobart," he said, "but I don't want you."

In this poisonous atmosphere, once again virtually isolated, Hobart buckled down to build the kind of armored division of which he had always dreamed. There was virtually no communication with main HQ, no sympathy with what he was doing, no cooperation and no equipment. Hobart proved his superb qualities under these negative, antagonistic conditions by bringing off the miracle of the 7th Armoured Division.

Troops accustomed to the sleepy garrison routine of Egypt found themselves with a stern task-master. Rushed into the desert to train by day and by night they soon found themselves permeated by the unconquerable spirit of the tall, hawk-faced Hobart. He infused them with the same magic morale he had given to the 1st Tank Brigade, and month-by-month he welded the scattered units into a determined, smoothly functioning fighting division.

Taking the jerboa (desert rat) as their emblem, they were soon known as the "Desert Rats." They proved themselves Britain's finest armored division in the whole North African campaign. Lieutenant-General Sir Richard O'Connor, commander of the Western Desert Force of 1940, called the 7th Armoured Division "the best trained division I have ever seen." [4]

The grim and frustrating duals of the War Office and the struggle for the armored idea slipped into the background as Hobart fulfilled himself in a man's job. When war broke out in September 1939, a deadly, hard-hitting and superbly mobile force was under his command. Lean, tanned and hard of body and mind, the 54-year-old Hobo was ready for whatever the war could bring.

Three months later, Hobart was dismissed from his command and sent into retirement.

This shocking blow came at the hands of General Archibald Wavell, who decided to act on an adverse report on Hobart filed by the general who hated him and who had sworn professional retribution. Normally a man impervious to the effects of opposition or professional misfortune, Hobart was shaken to the roots of his being by his abrupt and complete dismissal.

Lady Hobart recalls the 1940 dismissal from the army as the one time in their life together that the general had shown distress over any reverse. "He was a stricken man," she says today. "To anyone lacking his intense fortitude, the wound would have been mortal. No warning whatever was given that this blow was to fall."

General Sir Archibald Wavell, who was himself a man with a keen mobile sense, was unable in later years to explain adequately his action in dismissing Hobart. The loss of the tank genius from the desert command was to have incalculable consequences for British arms and fortunes. Liddell Hart tackled Wavell about Hobart's dismissal personally, and made it clear to him how deplorable and damaging the whole affair had been. "Wavell's explanation was rather lame," says Liddell Hart.

Wavell went on to win his own immortal glories by crushing the Italians with the Hobart-trained 7th Armoured Division – the only unit available and able to nullify the overwhelming Italian advantage in manpower and machines. By one of destiny's strangest twists, Liddell Hart had compiled a list of the most promising officers in the British Army for Hore-Belisha in 1937. Only two men were singled from the multitude of British generals as likely to become great commanders – Wavell and Hobart.

The fortunes of the British Army in North Africa were left after Hobart's dismissal in the hands of high commanders who were no more than amateurs in the handling of modern armored forces. So tight was the conservative grip on command that it was not until the latter part of 1942 that authentic tank officers even reached divisional commands. This continuing prejudice and incomprehension was reflected by the British Army's record in the field. With an inferiority of force but with an intuitive gift for handling mobile forces, Rommel proceeded to thrash humiliatingly a succession of British generals sent against him. The troops in the field, as well as the public all over the world, began to wonder if the British had ever heard of the tank before Rommel. British troops in North Africa, repeatedly let down by their armored forces, began to look on their own tank units with considerable suspicion.

When Hobart went back to England, an appeal against his dismissal was made to the king. The appeal was never put forward by the War Office. In Britain's time of mortal danger, Hobart's foes had eliminated him completely from military affairs, and had no intention of bringing his case to the attention of the monarch. For his general's uniform and badges of rank, Percy Hobart substituted the white brassard of the Home Guard on the sleeve of his lounge suit.

He joined the Home Guard without communicating anything of his intense disappointment to his wife and family. A deliberate effort had been made to break Hobart's spirit as well as to end his military career. Self-pity might easily have overwhelmed a lesser man, but Hobo was made of sterner stuff. "I cannot do what is ideal, so I must do what I can," he told his wife. He entered seriously into his Home Guard duties as a corporal. As the months passed, he seemed to develop an inner conviction that his chance would come, and that the wheels of the gods would eventually grind. For Hobo, the wheels of the gods ground along on German tank tracks.

Six months after Hobart's removal from the army, Guderian's panzers had run the British Army out of France in one of history's most humiliating routs. The able and farsighted German leader had used, to perfection in war, the techniques first tried and proved by Hobart. Never was there a more appropriate time for review of their military affairs and doctrines by the British, for only the miracle of Dunkirk had saved their beaten army from capture or annihilation.

Incredible as it must now seem, the stinging defeat of France and Dunkirk, with its devastating effects on morals and national pride, made little impression on Britain's military conservatives. Their intellectual detachment from the dynamism of events continued. The smashing of their First World War-type formations in France was deemed due to some sort of lucky German punch, even though Hobart's Tank Brigade exercises in the middle 1930s had portended the armored revolution with undeniable clarity.

Winston Churchill was not satisfied either with these military notions, or the defeats they had brought upon Britain. He was no friend of military die-hardism. One of the early pioneers of the tank in the First World War, Churchill had helped batter down opposition to its introduction into the earlier conflict. Between the wars, the future prime minister had watched tank developments closely. Hobart's disastrous misemployment incensed Churchill. As prime minister and minister of defense he was the most powerful official in Britain, but getting Britain's leading tank tactician and general back into the army was to take every ounce of his authority, as well as some of his eloquence.

As late as October 1940, Hobart was still unemployed, his appointment obstructed high in the War Office. Churchill was given a dossier listing the reasons why the progenitor of the *Blitzkrieg* should not be given an armed division. Churchill replied to the resisting spirits in the War Office with a historic minute. [5]

> October 19, 1940
> *Prime Minister to Chief of Imperial General Staff:*
> I was very pleased last week when you told me you proposed to give an armored division to General Hobart. I think very highly of this officer, and I am not at all impressed by the prejudices against him in curtain quarters. Such prejudices attach frequently to persons of strong personality and original

> view. In this case, General Hobart's views have been only too tragically borne out. The neglect by the General Staff even to devise proper patterns of tanks before the war has robbed us of all the fruits of this invention. These fruits have been reaped by the enemy, with terrible consequences. We should, therefore, remember that this was an officer who had the root of the matter in him, and also vision. I have carefully read your note to me, and the summary of the case for and against General Hobart. We are now at war, fighting for our lives, and we cannot afford to confine Army appointments to officers who have excited no hostile comment in their career. The catalog of General Hobart's qualities and defects might almost exactly be attributed to any of the great commanders of British history.
>
> ...This is a time to try men of force and vision, and not be confined exclusively to those who are judged thoroughly safe by conventional standards.

With this push from Churchill, Hobart's star went into the ascendant. He raised and trained the 11th Armoured Division, earmarked to fight in North Africa. While he set his indelible personal stamp on the 11th, Hobart chafed at the disasters inflicted on the British in North Africa by Rommel. He felt certain that he could defeat the Desert Fox if given the chance, but on the eve of the 11th Armoured's departure for Africa, Britain's military reactionaries took one last ignominious cut at the brilliant tank leader.

Prime Minister Churchill, left, accompanied by General Hobart, right, inspects the 11th Armoured Division in Nov., 1941.

Because his military views could no longer be gainsaid, the final efforts to oust Hobart were made on medical grounds, and mainly because he was now 56. His opponents were unfortunate in that they made their last efforts to ruin and remove Hobart in September of 1942, a black month for the British Army. Only three months earlier, Rommel had sent the powerful British 8th Army reeling back in a rabble from Tobruk. The Desert Fox stood now at El Alamein, readying his final thrust at Alexandria. This reverse had been inflicted by dynamically directed armored forces on the superior British Army and had left Churchill furious. The prime minister had also personally visited and inspected Hobart's new 11th Armoured Division only a few months previously, and had found Hobo in full vigor. Churchill's reaction to the final attempt to oust Hobart was this second historic minute on the tank leader, filed September 4, 1942: [6]

> *Prime Minister to Secretary of State for War:*
> I see nothing in these reports [of the Medical Board report on General Hobart] which would justify removing this officer from command of his division on its proceeding on active service.
> General Hobart bears a very high reputation, not only in the service, but in wide circles outside. He is a man of quite exceptional mental attainments, with great strength of character, and although he does not work easily with others, it is a great pity we do not have more of his like in the service. I have been shocked at the persecution to which he has been subjected. I am quite sure that if, when I had him transferred from a corporal in the Home Guard to the command of one of the new armored divisions, I had insisted instead on his controlling the whole of the tank developments, with a seat on the Army Council, many of the grievous errors from which we have suffered would not have been committed.
> The high commands of the Army are not a club. It is my duty... to make sure that exceptionally able men, even though not popular with their military contemporaries, are not prevented from giving their services to the Crown.

As it happened, the assignment of Hobart's 11th Armoured Division to North Africa was cancelled at the last minute. Under Major-General G.P.B. "Pip" Roberts, a Hobart-trained tank leader of great skill, the 11th later became Britain's finest armored division in the whole of the European campaign. Hobart raised and trained the two finest British armored divisions of the war, but a more massive challenge awaited him now, beside which an ordinary divisional command would have been misuse of his unique talents.

The invasion of Europe and the subsequent campaign into Germany required a host of new-type tanks armored vehicles. Tanks were needed for bridging ditches and rivers, clearing mine fields, throwing flame, destroying pillboxes and emplacements, and for swimming ashore from landing craft with the assault waves, and crossing rivers. Because these tanks did not exist in usable form, they had to be developed, together with the tactics for their employment. Men would have to be trained in the specialized task of manning these new weapons.

Design and development problems were enormous, and it was not a job for a riding instructor. Britain's new Chief of the Imperial General Staff, General Alan Brooke, had not been a Hobart enthusiast in prewar days. Nevertheless, he was man and soldier enough to recognize that at this juncture there was one man in Britain pre-eminently qualified to develop specialized armor for the invasion and conquest of Europe.

General Alan Brooke called a somewhat bewildered and cautious Hobart to his London office in March 1943 and asked him to train a unit in the handling of specialized armor. This unit was later to become known as the 79th (Experimental) Armoured Division. After almost two decades of frustration, disappointment, sidetracking and out-right victimization, Hobart suspected some sort of trap. Sir Alan Brooke's prewar apathy to the armored idea remained fresh in his mind. The ex-Home Guard corporal asked for time to consider the offer of command made to him by the Chief of the Imperial General Staff. Sir Alan Brooke agreed to this request, and Hobart set out to track down Liddell Hart and get his views on the proposal.

Hobart found Liddell Hart at the house of friends in Stoke Hammond, outside London. All urgency and energy, Hobo took the famed military analyst out in the garden for a private talk. Striding up and down in an icy wind for an hour, arguing about the new armored unit as a vehicle for Hobo's talents, they looked like anything but friends. Liddell Hart's wife, Kathleen, took periodic nervous looks out of the window. The vehemence of their discussion was unmistakable, and she wondered if they were quarreling.

Liddell Hart finally convinced the gun-shy Hobart that it was an opportunity to be seized, and that such a chance would never come his way again. The 79th was to be the biggest division in the world, and also the first all-armored division. Tempted by the prospects, excited by the challenge, Hobo's resistance crumbled. He took the job.

Hobart's drive, knowledge and willpower became decisive in the building of the epic 79th. Time was short. There was virtually no background of previous experience on which to draw, a situation which placed a premium on Hobart's acumen, experience and military intuition. Challenge and fulfillment came together.

Trials and tests were endless. Hobart's gift for arousing enthusiasm for a new idea found full scope. The 79th (Experimental) Armoured Division took a bull's head as its insignia and soon boasted the same kind of soaring élan and confident professionalism that characterized other Hobart-trained formations.

Urgency and excitement pervaded Hobart's environment, and no longer were there blockheads in brass hats to scrutinize and obstruct his requirements. On the contrary, men with wide authority moved heaven and earth to provide him with the necessary resources.

Field Marshal Montgomery, the conqueror of Rommel, was Percy Hobart's brother-in-law. Although a Hobart admirer for many years, Monty had tended to shy away from the tank idea when it was unpopular at the War Office. The hero of El Alamein now put his prestige behind Hobart's work and took up the needs of the 79th with General Eisenhower. The Supreme Commander quickly recognized Hobart's vital role and his unique abilities in developing specialized armor. Eisenhower slashed red tape and gave top priority to the US manufacturer of the odd-looking tanks and attachments Hobart required. High-level push of the kind, and Eisenhower's unstinting support of anything likely to save lives, soon provided the resources to assemble Hobart's "Menagerie," as it became known.

Liddell Hart has called the 79th Armoured Division "the tactical key to victory." Because it was not a division that fought as a unit, but had its elements farmed out to the Allied armies wherever they were needed, the 79th has far less historical fame than most of the Allied divisions that stormed through Europe. How far many other divisions would have been successful without the "funnies" of the 79th is a question for debate.

By the time the Allies reached the Rhine, Hobart's 79th division consisted of eight brigades and a total of 17 regiments, quadrupling the complement of armored and tracked vehicles on the establishment of any normal armored division. This huge metal menagerie was spread out at times over a front of ninety miles, and the direction and allocation of its 1,900 armored vehicles kept Hobo hopping.

As the US Army in the beginning did not have specialized armor of its own, the 79th frequently worked in close support of US troops, and was the only British unit to do so. This situation suited Hobart. He liked Americans and they liked him. He was direct, frank and forceful, knew what he was talking about and understood the American character as few British commanders ever did. He would verbally thrash any officer or man he heard speaking against the Anglo-American alliance, to which he was deeply devoted. At one time, he even had an American aide, New York oilman George Thomson, Jr., who served with the British army. Hobart's radiant admiration for things American, such as know-how and mechanical skill, was not a superficial or transitory thing. He had an intimate knowledge of American commanders and their views, and an extensive knowledge of US military history. He held America's top generals in the highest regard.

The directness and honesty of most American generals appealed greatly to Hobart. With the US 9th Army commander, General W.H. "Big Bill" Simpson, the feeling was mutual. Simpson was taken aback by Hobo's quiet boast that he was "the oldest major-general serving in Europe." Simpson says of the amazing Englishman: "He was the most outstanding British officer of high rank that I met during the war, and from his mind and bearing no one could possibly have guessed his age."

Vigorous and vitally alive, Hobart served with his fantastic steel menagerie until the final gun of the war from which he had almost been excluded. The case for armor had been proved. The basis for future manifold developments of tanks had been laid by the accomplishments of tanks had been laid by the accomplishments of the 79th. Wrote General Eisenhower in his report: [7]

> Apart from the factor of tactical surprise, the comparatively light casualties which we sustained on all beaches, except OMAHA, were in large measure due to the success of the novel mechanical contrivances which we employed, and to the staggering moral and material effort of the mass of armor landed in the leading waves of the assault. It is doubtful if the assault forces could have firmly established themselves without the assistance of these weapons.

Hobart had probably done more than any other single individual to advance both tanks and specialized armor on the practical level. Had Hobart's 79th Armoured Division, with its fearsome bull's head insignia, not been such a spectacular success, tank innovations may well have halted as they did after the First World War. Tanks are today an integral part of atomic battlefield planning.

Percy Hobart was knighted by King George VI, and from the US received the Legion of Merit, Degree of Commander, a decoration of which he was extremely proud. When he went into retirement after the Second World War, it was in an honorable and upright way, with his admirers far outnumbering his critics. His death in 1957 saw him deeply honored and widely mourned, and to have "served with Hobo" is a real distinction in the British Army, where his one-time juniors and students are now in the highest commands.

From persecution, victimization, and his incredible misemployment as a Home Guard corporal, Hobart's resurrection to a decisive command in the Allied armies is one of the more startling personal stories of the Second World War. His story was hardly the kind of thing likely to impress the public with the efficiency of the war effort, or the quality of Britain's military leadership. Thus he remained almost unknown outside army circles.

The most memorable tribute to Hobart came from Captain B. H. Liddell Hart, whose exposure of the Home Guard episode started the tank pioneer on the road back. All the high British commanders and most of the Americans had passed before the famed analyst in a living parade, as they pursued their careers and often aroused his criticism. Liddell Hart also knew the Germans well – perhaps better than any other military writer and thinker outside of Germany. As Britain's leading military brain, his judgment has many times been vindicated, although his warnings all too often went unheeded.

In Liddell Hart's opinion, the independence of a top command would probably have proved Hobart to be the best of the British commanders, capable of matching the best of the Germans on equal terms. In summing up, Liddell Hart writes of Hobart: "He was one of the few soldiers I have known who could be rightly termed a military genius."

Notes

1. Personal reminiscence provided by General Sir John Crocker, Hobart's brigade-major in 1934.
2. Personal remembrance of General Sir John Crocker.
3. Personal recollection of Lady Dorothea Hobart.
4. General Sir Richard O'Connor, commander of the Western Desert Force, 1940-41. Cited in: B. H. Liddell Hart, *The Tanks* (Praeger, 1959), vol. I, p. 404.
5. Winston S. Churchill, *The Second World War, Vol. II*, "Their Finest Hour," 1st ed., pp. 602-603.
6. Winston S Churchill, *The Second World War, Vol. IV*, "The Hinge of Fate," 1st ed., p. 791.
7. Cited in: B. H. Liddell Hart, *The Tanks* (Praeger, 1959), vol. II, p. 332.

Author's Note

This article is slightly adapted from a chapter of my book *Hidden Heroes*, which was published by Arthur Baker, Ltd. Since then, this unique collection of biographical sketches has received no exposure or publicity.

Consequently, the little-known Second World War tale of Percy Hobart's victimization and vindication is presented here, for the first time ever, to an American readership.

I remain much obliged, even after more than 30 years, to the late eminent military historian and analyst, Captain Sir Basil H. Liddell Hart. He gave freely of his professional time to assist me with numerous details, insights and clarifications. He patiently corrected my drafts of this story, in which he himself had been intimately involved from first to last.

The late General William H. Simpson, former commander of the US Ninth Army, enthusiastically shared his reminiscences of General Hobart. George Thomson, Jr. of New York, and Major John Borthwick of Britain, military aides to General Hobart after his "resurrection," provided valuable insights, each from his own perspective, into a many-sided military genius.

The late Generals Sir John Crocker and Sir Harold "Pete" Pyman, similarly contributed to this portrait of Hobart, as former students who lived not only to see their visionary teacher's predications come true, but to be developed further in scarcely conceivable ways. Lady Dorothea Hobart, the great man's widow, rendered indispensable aid by rallying these eminent men to help me, and was throughout the soul of kindness.

Liddell Hart on Hobart

"Much of the credit [for the February 1941 British victory against larger Italian forces at Beda Fomm, Libya] was due to a man who took no part in the campaign – Major-General P.C.S. Hobart, who had been appointed to command the armored division in Egypt when it was originally formed in 1938, and had developed its high pitch of maneuvering ability. But his ideas of how an armored force should be handled, and what it could achieve when operating in strategical independence of orthodox forces, had been contrary to the views of more conservative superiors. His 'heresy,' coupled with an uncompromising attitude, had led to his removal from command in the autumn of 1939 – six months before the German panzer forces, applying the same ideas, proved their practicability."

—B. H. Liddell Hart, in his *History of the Second World War* (New York: 1971), p. 117.

Chapter 12

THE MASK OF OFFICIALDOM

Excerpt from *The Cosmic Pulse of Life*

One must live things to judge them. —Wilhelm Reich

Many people were prompted to try and help me after publication of *They Live in the Sky* in 1958. Interested and kind people who already knew me, and many who sought me out in a spirit of intelligent good will, did their best to "promote" me and my work – despite its underfinanced primitiveness. Introducing me to scientists and engineers of their acquaintance was their major effort on my behalf. They were rational people, but unacquainted with the formidable psychological problems attendant on the discovery of biological UFOs, and therefore these efforts all proved unrealistic and fruitless.

The rational aspect of the efforts made to help me was based on what people could observe happening in the U.S. The Pentagon was mindlessly dumping scores of millions of dollars into worthless "make-work" research in many fields. An influential bagman in Washington seemed to be all that was needed, once you had a covey of researchers with official credentials on the payroll or otherwise on tap. Introducing me to steady, established, thoroughly safe individuals was the procedure my friends thought would be most fruitful.

All the well-intentioned efforts made to assist me were barren of results in any immediately applicable fashion. There was nevertheless an invaluable and unforeseen dividend. The efforts involved me in numerous meetings with a variety of scientists and engineers. As a participant once again, I learned for myself at first hand about the sham of so-called scientific objectivity.

Wherever scientific objectivity may turn up in the universe, it is most assuredly not of this world. My experiences with human irrationality in connection with radical findings are of minor dimension. Suffice it to say that the persecution of pioneers in cosmic electronics like Dr. Albert Abrams and Dr. Ruth B. Drown, and revolutionary cosmologists like Dr. Wilhelm Reich and Dr. Immanuel Velikovsky, verifies that this irrationalism is amplified to the dimension of the personality it becomes necessary to execrate. The ultimate manifestation of irrationality becomes the murder of the innovator.

By mid-1958, Jim Woods and I had accumulated well over one hundred mutually corroborative photographs of the two basic types of UFOs – vehicles and critters. Motion pictures of the critters whirling around me in the desert dawn decisively crushed the objections raised to the stills by self-styled experts in whom these strange critters seemed to evoke an unbearable anxiety. Seen in its totality, with the underlying reasoning and basic explanation of the experimental methods presented in this book, the collection of photographs made an overpowering impression on all who came with open minds.

There was no question that it all lay outside the boundaries of official science, and violated in the mode of its acquisition all the principles set up by the reception systems for the dissemination of scientific knowledge. Scientific periodicals returned contributions as though from a slingshot. We had violated most of the rules of mechanistic method, and one rule in particular, for which we would always be rejected by official circles. We had involved ourselves in the total process in a deliberate, outrageous and unconventional fashion.

Physics sometimes terms the influence that the individual experimenter has on the outcome of his experiment "the Heisenberg effect." This term commemorates the late Professor W. W. Heisenberg, who drew attention to this phenomenon. In most experiments in mechanistic physics, the Heisenberg effect is small enough to be ignored.

In our work, our sense apparatus was used in an extended, new way, impermissible to a mode of scientific cognition based on evidence received through a single, color-blind eye. The human body force field, or orgone energy field, was an integral part of our work, and success depended upon its methodical manipulation. The total human being was an essential part of the whole happening, and not an incidental effect influencing the results only slightly. To the orthodox mind, we were simply out of sight.

Without an understanding of the marvelous reality of the human visual ray, and direct contact with the living energy ocean in which all these UFO events were occurring, we would never have been able to produce any objective evidence. To this day, no one may say with certainty to what extent my particular personality and extra-physical energies were responsible for the results, since no one else has tackled the problem in the same way. A few people tried, got results, then became frightened, for one reason or another. No one persevered.

Each human is the bearer, at the subconscious level, of a variety of powers that transcend current scientific knowledge. The New Knowledge leaves no doubt that each of us is not a new soul – a brand new entity coming by accident into a world of indeterminate origin – but rather an entity thousands of years old. Given the right external conditions and opportunities, the things we knew and could do in other lifetimes and other lands manifest again as talents, drives and insights.

Our penetration into a supersensible stratum of physical nature invariably proved to be too much for scientists and engineers to whom I was directed by my well-meaning friends. Brainwashing by the universities, the systematic biasing of young minds to a mechanistic world order and an irrational financial system, is the biggest single barrier existing today to the introduction of the New Knowledge. Every time I went to people who were products of this system – at the behest of others – the experience was futile and frustrating.

Where qualified people came to me on their own, impelled by their own forces and interests, the results were quite different. Friendships were established that have endured ever since. When these people of novel bent broached these matters to fellow scientists, their experience proved to be the same as mine. They were beaten down by a peculiar, highly emotional and irrational reaction that bore no relevance to the facts being discussed.

Intolerance is the correct description of this reaction. After detailed study of Dr. Wilhelm Reich's work many years later, I came to understand what upset these people and why. The intolerance is the reaction of a neurotic personality to his own bio-energetic movement. Neurosis is born of chronic bio-energetic inhibition. When the individual's bio-energy moves in response to stimuli that are lawfully rooted in basic life processes – in this case the living creatures of our atmosphere – the neurotic individual clamps down on this movement. He literally cannot stand it.

Such reactions were observed by me for years, and some of the most highly qualified men I talked to reacted most violently. These reactions were utterly incomprehensible to me at the time because I had been brainwashed with the legend of scientific objectivity in my own education. There is no more fatuous myth permeating our modern life than the one that tells of detached, unemotional men of science. I have seen Life scare them stiff!

When we deal with the scientific work of Wilhelm Reich in later chapters, all this will be further elucidated. Having seen this queer reaction scores of times, among all kinds of people whose views on extraterrestrial life are already concreted within the mechanistic framework, I believe it to be an integral part of contemporary human attitudes and reactions to life in space. Ash cans to the moon at $30 billion per voyage do not move the bio-energy. Critters from our atmosphere do. People can tolerate the former, but not the latter.

Persons who can tolerate new work of this kind, usually have only one basic question if they are without technical expertise: 'What do the physicists say about this?" Reference is automatically and always to some form of authority, even though such authority may have no relevance to this work, which is new and revolutionary.

There are no authorities on this work, not even me.

Almost two decades after I started, I know only that my ignorance of what lies behind and beyond the realm into which I have broken is appalling. Pandora's Box was by comparison a picnic basket. Experience has taught me that the most brilliant men are, from the cosmic point of view, still only on the fringe of understanding. Scientific cognition began for man only yesterday.

In the 1950s I was an idealist about the world of science, not then having learned to distinguish between science as an investigative instrument and method – which is marvelous – and the character structures of the individual human beings who turn science into scientism. Because of my naive idealism, I was persuaded to seek the approval of physicists, chemists and others who were qualified in their own formal fields. These people had either assumed authority in the UFO field, or had authority thrust upon them by laymen who deferred to them in technical matters. This assumption or assignment of authority had persisted despite three decades of zero progress by formal science in the UFO field, and the unavoidable conclusion that something is lacking in mechanistic scientific cognition.

Irrational reactions toward New Knowledge are not confined to my work on the part of the conventionally minded. Highly qualified scientists who have broken into the New Knowledge invariably have the same problem. One might cite here the work of Dr. Harold Saxton Burr with his colleague Dr. L. J. Ravitz at the Yale University Medical School.[1] They discovered the Life-fields of which everything discussed in this book is in some way a manifestation – including the critters. Prejudice and resistance attended this epochal work, although it was done in an august institution of higher learning.

The UFO project at Colorado University demonstrated little concerning UFOs, but was rich in lessons concerning human beings. The characterological barriers to straight mechanistic investigation of UFOs were fully exhibited. The project also showed that the Establishment can buy convenient verdicts from teams of scientists – which is corrupt and wrong by any normative standards. Genuine scientists resist corruption, and it was such men within the Colorado project that triggered the controversy that erupted.

The Colorado project was staffed with scientists qualified in their disciplines at the best universities. Men who had studied UFOs avocationally were ruled out of the teams, so that experience, familiarity with the subjects and any insights accordingly won might not contaminate the pristine product the Establishment had ordained should come forth. The pasteurized, thoroughly safe men who gathered at Colorado were to scrutinize and evaluate UFO evidence.

Soon they were accusing each other of duplicity and engaging in squalid quarrels through books, magazine articles and other media. There were firings of top personnel and naked threats of professional vengeance made against those scientists who declined to see the Colorado project run quietly along the rails laid down for it by the Establishment. The largest civilian UFO group in the U.S.A., Donald Keyhoe's NICAP organization, withdrew its support when evidence NICAP made available to the project was not properly utilized.

For years Major Keyhoe had called for a government UFO project, staffed with top scientists, as a necessary and desirable official approach. Numerous other UFO investigators, with similar high expectations from scientific objectivity, similarly urged just such an official approach. When at last these importunings were answered, the character structures of the participating scientists made a public shambles of their objectivity. Whether or not the individuals involved could bio-energetically tolerate UFOs seemed to determine which of the two camps they joined. The anti-Life faction won control.

UFOs are pervaded with life. At Colorado, some people could tolerate life and movement, including their own corresponding organ sensations. Other people were anti-Life, or Life-negative. Any well-authenticated body of UFO evidence has life in some way immanent in its substance. This is true even if one can stretch no further than the seedy extraterrestrial intelligence hypothesis. Even this is life of a new kind. Some neurotics find this highly disturbing, and the disturbance makes them act irrationally.

Some of the Colorado scientists stood up for Life. The rage and threats against them, well-documented in the whole lamentable episode, sprang from frightened, irrational men, their scientific detachment blown to the winds by their terror of their own biological movement. What happened to the honest and upright scientists at Colorado had happened to me many times and many years prior to Dr. Condon's project.

Having experienced this irrationalism among scientists at first hand over a long period, I was able to predict the outcome of the Colorado project with precision. Barely a month after the USAF announced the contract, I wrote an article for the *Journal of Borderland Research*, predicting that Emotional Plague would break out despite the august university atmosphere. I said that nothing would come of it but chatter and bewilderment. The Journal's editor at that time, my good friend Riley Crabb, was jam-packed with articles and did not use my piece, but later said he regretted not having scooped other periodicals. In the UFO field, most publications opined that the scientific approach would win out. So it will, when we understand and reform human character. My negative advance judgment on the Colorado project was based on bitter experience with a wide variety of troglodytes.

In a typical instance, a close friend of mine who was an independent UFO researcher, persuaded me to meet with a group of scientists at a suburban home in Los Angeles. I will call the hosting engineer George Broughton, although that is not his real name. He has always maintained his goodwill toward me, and was not responsible for what happened. His interest in UFOs hurt him economically in many ways, and I refrain from identifying him because of his Federal business connections and obligations.

Broughton had an electrical engineering degree and a sizeable plant doing defense work. UFOs intrigued and enthralled him. He was fascinated by my photographs and the possibilities they opened. He could also see the commercial future in mastering the energy form that propelled the discs. The biological UFOs he took in his stride. "The photos speak for themselves. They eliminate a hell of a lot of confusion about these things," he said. Broughton was a straight, bright, unblocked individual who could tolerate Life and the New Knowledge, even if the latter had not yet been brought down into hardware.

The men he invited to his home that evening were of a different stripe. So was his wife. She greeted me with ill-concealed hostility and abruptness, embarrassing him and establishing an antagonistic atmosphere. She hated UFOs passionately, and hated her husband's interest in them. To the kindly Broughton's dismay, the proceedings were taken over by a bumptious biologist, who said to me before we were even introduced, "Has any of this crap you're putting about been proven?" Without apprising Broughton, the six men present had evidently agreed to give me an uncomfortable evening – to put me down.

When I tried to give a basic account of the methods employed to attract UFOs and photograph them, the biologist kept interrupting with "But you can't do that!," dragging me into arguments on side issues and quibbling over methodology. My host protested in vain. Two other men present, both aerospace industry engineers, hectored me relentlessly.

When the time came, I opened the case in which I carried eighteen enlargements of my best photographs. I handed the "amoeba" photograph shown in Alpha #1 to the biologist, and held four successively captured shots of the same critter in my hands. He looked as the "amoeba" and immediately looked away as though needles had shot out of the picture into his eyes. He handed the photograph back to me quickly. "I don't see anything in this picture," he said. I invited him to look at the sequence of the pictures, so he could follow the obvious expansion of the object, but he flatly refused. "No true scientist would ever look at such stuff," he said.

Years later, I understood what Dr. Velikovsky encountered in his struggle for a new cosmology, when eminent, internationally famous men of science put him down in scientific journals. The tirades against Dr. Velikovsky were characterized by the writers stating that they had not read his *Worlds in Collision*. Not to read Velikovsky was an article of faith among his critics. The scientific world is riddled with, and addled by, such irrationalism.

In two grueling hours with the six men at Broughton's house, not more than a dozen intelligent questions were asked. Most such questions came from Broughton. The other men all spoke quickly, with an evident need for reassurance from the others. In those days, I did not understand what was happening in a behavioral sense to those men. I did understand that objectivity, calmness, fairness and rationalism were not there. Something bordering on panic was present in that room. Normally calm and objective men were experiencing anxiety.

Striking indeed was the compartmentalization of their knowledge. All of them were sort of walled off from each other. The electrical engineer grasped the photo evidence of quasi-electrical propulsion, but dropped out in dealing with anything obviously biological. The physicist understood the plasma effects around some of the objects, but recoiled from even the tentative idea that organisms and vehicles were somehow functionally related. All of them were basically kind, normal men, husbands and fathers, well-educated and respected, but in terms of both scientific training and character structure they were not qualified to deal with what I had uncovered. They were tragically emblematic of the flight to scientism when the New Knowledge breaks through.

They were a microcosm of the whole scientific failure and default on UFOs. Getting these intelligent men to accept that I had photographed these things when they were invisible proved impossible. In that respect, they were all totally blocked. When I went home, I was deeply troubled by the multiple examples of irrationality that had given me one of the most uncomfortable evenings of my life.

How could men of science be so emotional, so intolerant, and so strangely fearful? That question arose again and again. Why the antagonistic reception to something innovative – something that was throwing light on the greatest mystery of all time? Until these experiences with workaday scientists crowded in on me, I had always felt that science was a pure instrumentality, untainted by emotion. At first hand, I was learning a central fact of 20th century life: science can never be any better than the character structure of those who labor in its service. If the scientist's character structure is mechanistic, so will be his perceptions, reactions and conceptions. Their direction can only be toward the crushing out of Life.

Next day, Broughton called me and apologized for his guests. We had lunch, and he told me that after my departure, his friends ridiculed him for his support of me. He made light of it at the time, but had been hurt by their behavior. He told me that in his opinion I was too far ahead of my time, and too mercurial for orthodox scientists to follow. Ten years later, I met him by chance in the Mandarin Hotel in Hong Kong, and his first words were, "I'll never forget that awful night at my home…" Nor will I.

Another similar incident illustrative of orthodox reaction to work like mine, took place soon afterward. A close friend of mine, graduate of one of the finest U.S. universities and well-connected in the Defense Department, persuaded me to try and show my work somewhere in official quarters. Earlier I had placed everything at the disposal of the USAF, but there was simply no interest in such work, even though the USAF was dumping money into so-called "think tanks," wherein bright boys were paid to speculate about spacecraft propulsion.

A negative reaction from the Federal government does not necessarily mean that what one is offering is no good. The defense services in all countries are markedly resistant to new concepts of all kinds, even when they originate with responsible and qualified people. Innovators who are too daring and too forceful have to be torpedoed, as was the case in the U.S. with General "Billy" Mitchell. Irrational opposition from high-ranking professional officers to new conceptions is a central element in military history.

Each world war has proved that only a handful among the hordes of professionals is able to handle high command successfully. People still trust the military professionals despite their repeated failures even to provide the proper patterns of weapons before major conflicts erupt. My service-connected friend suffered from overconfidence in the defense establishment. He made the arrangements for me to visit a Defense Department research office in Los Angeles to show my films and photographs. Still naive and idealistic, I felt that with his introduction there would be no repetition of that awful affair at Broughton's. After all, I told myself, that was essentially a social affair, and this was a business visit to a Federal office.

The physicist who received me opened our conversation by stating that the official position was that these things do not exist. When I sought to explain the simple methods used to obtain the films and photographs, the physicist kept looking at his watch. He had no interest in a specially prepared folder of fifty photographs I had prepared for this day. In my galloping naivete I offered to leave him my negatives for analysis, but he said it was not necessary.

The motion picture films were disposed of by an intelligence officer, who then projected them upon a dirty and pockmarked wall, winding the lens in and out of focus as the film was shown. Despite this chicanery, sufficient footage came through clearly – dirty wall notwithstanding – for it to be a convincing introduction. The physicist left the room, red-faced and angry. The intelligence officer handed me back the film. "Now you can say, if you are asked, that this office has seen your film and pictures." He then walked away.

Persons who doubt that such reactions occur, should undertake some serious time to the advocacy of any pro-Life viewpoint. Thanks to the findings and discoveries of Dr. Wilhelm Reich, the etiology of the physicist's irrational fury is today well recognized in orgonomic psychiatry. Today's ecology advocates are encountering kindred irrationalities, in their efforts to prevent man from suffocating himself in his own effluvia. To me, in those days, as a young man unlearned in the ways of neurotic reaction, I was thunderstruck that a man who had earned a Ph. D. degree would blow his mind and temper over a piece of scientific film.

These reactions typify untold dozens that I experienced through the years. Whenever I made the mistake of approaching some established, orthodox-minded person, usually at the behest of a well-meaning third party, I could depend on such reactions – in greater or lesser measure. There was also the studied vacuity of official, government reaction, as a counterpart to the individual emotional volcano.

The late Senator Clair Engle of California was a licensed private pilot, and also had a strong personal interest in UFOs. We had a brief correspondence, as a result of which he sent some of my material to the USAF. The reply which came from the Air Force credited me with having photographed "space ectoplasm," a term never used and a claim never made by me. The USAF view was that the things I had photographed had nothing to do with the UFO phenomena. And having made the pronouncement, the matter was dropped without further communication or inquiry into what would nevertheless be an amazing discovery in space, even if it didn't have anything to do with UFOs. Let all others who photograph pulsating, glowing discoidal forms be hereby warned.

The dynamic activity of the dawn period, when the bulk of my early photographs and films were made, is not something invented by me. Rather it is a phenomenon independently observed by scientists, as a result of radar propagation anomalies that have been observed and recorded at this time of day. In layman's terms, a "propagation anomaly" is an irregularity in the normal way in which radar signals are transmitted and received.

One of the anomalies that bothered radar scientists in the late 1950s and early 60s was the reception of strong radar returns from objects that are not ocularly perceptible. "Angels" was the term applied to these objects that seemingly are not there, except that radar says they are. My photographs are of objects that are also not seen with the naked eye – except sporadically and occasionally – at which time they become UFOs.

An Air Force project conducted by the Cambridge Research Laboratories attempted to explain why service radars detected strange clumps of targets in the pre-dawn period and also before sunup. The findings, theories, conjectures and speculations of the baffled scientists were presented by Vernon G. Plank in two papers published by the Geophysics Research Directorate of the Air Force Cambridge Research Center. Paper No. 52 was "*A Meteorological Study of Radar Angels*," and Paper No. 62 was "*Spurious Echoes on Radar, a Survey*."

Anyone who thinks that my finding invisible flying objects in the earth's atmosphere at dawn, and photographing them, is farfetched should read the alternative explanations in these two enlightening treatises. Imaginative explanations for the dawn "angels" are offered that are far wilder than mine, and for these explanations no proof is adduced. Flocks of birds, heated pockets of gas, clouds of insects and other way-out notions were offered as examples of what the angels might be.

My friend Bob Beck, a former engineering test pilot and an instrumentation specialist, read the monographs and expressed my feelings precisely when he returned them to me.

" Those guys," he said, "are really reaching."

Some young scientists who had been concerned tangentially with the angels project, and who knew that they were "reaching," were shown some of my photographs by an electromagnetic interference expert who had an interest in my work. These young men could see a promising and mighty solid explanation for the angels that they had been dogging.

True to the spirit of science, they wanted to corroborate my findings with more orthodox methods and apparatus. I was ready to assist in any way open to me, in complete anonymity and without any desire to receive official credit, funds or recognition. For me, it was enough that these young men should be able to pull off the corroboration of what I had done.

When this proposal got up to the level where heavyweight scientists hand out funds and approval, the whole idea was killed. That life might be present in a heretofore unsuspected form was evidently once again going beyond what the older men in science could tolerate. A couple of ardent young spirits wanted to pursue the venture on their own time with their own funds, but were denied use of service apparatus and told to drop the approach completely. The angels continue their anomalous manifestation, but it is enough to give them a name – angels – and leave explanations for later generations. If it is alive, then don't touch it.

There were many other instances involving scientists and engineers who tried to get attention to my work. Hope was always held out, despite the radical nature of my findings, that a more conventional empirical approach might verify my findings in an acceptable fashion. Nothing ever came of all this pulling and hauling. My personal policy was to greet all aid with goodwill and high hope, always assist to the maximum, and to stand aside completely if someone should insist on personal credit.

Every individual who tried to work within any kind of official framework, either in government or in private industry, got set down hard by his superiors when he broached biological UFOs. They experienced frequently the same Emotional Plague reaction that I have already described. In every case, they were men with families to support, and were honest, open-minded and fair men. The threat of economic reprisal – open or tacit – always caused them to recoil, and for that they cannot be blamed. In a hardware culture, living needs are necessarily secondary to the primary mission of selling junk to the government. All these well-motivated efforts were thus either stillborn or strangled at birth.

These examples paint a general picture of scientific reaction to this approach to UFOs, when it was advanced in the 1958 period and in the years immediately following. Considering that the publications, bodies and individuals approached were essentially products of and enslaved by the mechanistic world conception, the reaction is really not surprising in retrospect. Among genuinely friendly, open-minded people of science with a feeling for the novel, however, the reaction was not greatly different. Their cutoff point simply came later.

A close friend of mine enjoys the friendship and confidence of many of the top men in science in the eastern U.S. Many of them are world famous for their discoveries, inventions and achievements. They meet periodically to hash over the psychic, the occult, the new and the way-out, seeking stimulus and diversion from the workaday concepts and methods by which they make their livelihoods. All of them were fascinated by my photographs, which continued to accumulate into the Bravo Series, some of which are shown in this book. The Bravo Series includes a remarkable set showing USAF jet fighters, armed with Sidewinder infrared homing rockets, chasing my critters above me over the Mojave Desert.

These men could tolerate the photographs and the findings. They were delighted at the unequivocal proof that the USAF has lied about its pursuits of UFOs. Helping the work, or introducing it outside their secret circle, was something else.

They quickly joined the ranks of the "if only" men of my experience. "If only you had taken this in stereo, it would be incontrovertible," said one famous man. When stereo photographs were made, there were new evasions for new reasons, not the least of them that the objects appeared to be in two different places at the same time. The new problems were thus thornier than the old, something to which biologist Ivan Sanderson has referred in his account of my work in his book Uninvited Visitors.

The package could not be made neat and unobjectionable. Disturbance or destruction of the illusion of neurotic security under which scientists labor – no matter what their eminence – was simply unavoidable. The old criteria of reality were under fire. As greater ramifications appeared out of my efforts to shape the phenomena to what the mechanistic mentality would tolerate, my need for assistance rose sharply.

Revolution leaped out of the photographs. There was no way I could soften or lessen its impact. All the men of novel bent and goodwill eventually quailed before the challenge. I ended up with friends, sympathizers and even admirers, but without the support I needed. Being told I was fifty years ahead of my time was no consolation.

Fairness and chivalry constrain an understanding of these reactions among active scientists. Their professional tribalisms have not only form, but also force. Families have to be supported and responsibilities discharged that require maintenance of professional integrity and standards, regardless of the fascination a scientist may have for off-the-track findings and borderland phenomena. Retribution attends such dabblings if they become too vigorous. This kind of intellectual and economic blackmail is largely responsible for the general failure of organized science to tackle UFOs as serious business. Only hardy mavericks buck the system. Borderland roustabouts like myself cannot hope to alter the course of such a high-powered juggernaut.

The same kind of hedging and resistance extends into ufology, a subject where involved parties like to think of themselves as being open-minded. Ufology has nevertheless developed its own tribalisms. Official ufology appears to have no intention of departing from its prime dependence upon the ETH, or extraterrestrial hypothesis. While reasonable and probably in some ways true, the ETH is nevertheless inadequate and is unproved to this day.

Opposition to the New Knowledge approaches to UFOs takes the form of evasion of anything not readily classifiable as a spaceship. The more "solid" such a ship may appear, the more solid the attention directed to the incident. That unseen intelligences had given me basic information that I had used to obtain my photographs – to bear into a dynamic borderland of energy and force – was used as sufficient reason to discount my findings.

The irrationality of this posture was never apparent to diehard ETH advocates. Extraterrestrial spaceships were here, according to them. All evidence, in their view, pointed to this. Yet when a biological communications system provided information and methods for making these photographs, originating with entities who stated that they rode in ships from other dimensions, official ufology could not tolerate the new realities.

Ufology also hid behind the mask of officialdom, croaking down through the years for a government investigation of UFOs. When the investigation came, the qualified scientists who ran it turned more than half a million dollars of Federal money to crushing ufology. Twenty-first century students of 20th-century mass psychology will have mordant comments on the way fear of life powered this irrationalism. The prized, cherished, adored and worshipped objectivity does not exist among mortals.

Only a few periodicals remained opened to my contributions on UFOs, and beyond these it was as though an invisible wall had been erected against dissemination of my views. Editors of general interest magazines and men's magazines, to whom I was steadily selling military articles and stories, quickly fired back UFO pieces as "too wild," or routed them immediately to a tame consulting scientist for the death blow.

One compiler of a book of UFO photographs was told by a young lady who knew of my work that he should include something of mine. At her behest, I went to see him with a satchelful of UFO photographs. These pictures showed a wide range of forms and shapes of unseen things in the air. He already had many of these same forms in sketches provided by persons who had made sightings.

These sketches were on the desk in front of us, in paste-ups for the book he was compiling. In any other field than ufology, my photographs would have been like a gift from heaven for anyone compiling such a book. They objectified photographically what independent observers had already sketched, from sightings made in many parts of the world.

Like the biologist at George Broughton's home and dozens of other people in the interim, the book editor looked at the photographs quickly and then averted his gaze, like a Milwaukee matron finding a French postcard. He handled the piles of pictures as though they were physically hot. He couldn't stand the sight of them. He was uncomfortable and anxious in the presence of this pile of proof that fitted in so well with what he planned to publish. Finally he got it out.

" These are very interesting," he said, "but of course, they don't have anything to do with flying saucers."

As he spoke those words, the discoidal, glowing shape of the amoeba stood on the top of the stack. This incident more than any other, coming as it did after years of similar reactions where the associations of everything were less direct, convinced me that there is something fundamentally wrong with the way a great many human beings perceive. This man was obviously blocked. So were all the others who had acted irrationally.

By contrast, I noticed the way in which free spirits tackled the examination of the photographs. There was no looking away. They went right into the pictures with their gaze, pored over them, and inevitably compared them with microscopic forms. People like these seemed to be more alive and direct than those who couldn't stand to look at the photographs, and I always seemed to have a deep and immediate contact with them.

Official science and official ufology still wear the same mask to this day, although their irrationalism is no longer the enigma to me that it was in earlier times. The etiology of this irrationalism is understandable – and therefore ultimately conquerable – because my individual pathway led me to three giants of the New Knowledge. All three of these titans encountered similar irrationalism and rage in making their contributions to human advancement.

From these three titans I learned in fortuitous sequence the things I needed to know in order to understand something of what I had stumbled upon. I also learned how to carry it forward. The three avatars were Dr. Rudolf Steiner, Dr. Ruth B. Drown and Dr. Wilhelm Reich. Their work, discoveries and inventions permit us to begin a new, Life-positive science. Through Steiner in particular, we can understand – and therefore counteract – the hidden powers from beneath man, whose earthly works include the truth-killing mask of officialdom. Let us meet these three titans in turn, in the same sequence as I did, and we will begin to understand the Battle for the Earth that has already begun.

NOTES

1. See Fields of Life, by Dr. Harold Saxton Burr, Ballantine Books, N.Y., 1973.

Chapter 13

RAIN ENGINEERING QUESTIONS & ANSWERS

Introductory Statement

Planetary drought problems are advancing relentlessly, with their losses and tragedies and colossal waste. Countering desertification at source is beyond conventional, orthodox capabilities, which are essentially limited to mitigating the consequences of drought and managing water resources. As a crisis scenario unfolds worldwide, every promising new approach to increasing and targeting rainfall should be tried, and judged exclusively by *results*.

E.R.E. is a Singapore corporation formed specifically for the commercial implementation of airborne etheric rain engineering. My personal experience with this environmentally pure technical development exceeds 13 years. Association with Trevor James Constable has involved me personally in rain engineering operations, not only in a business capacity, but also as a hands-on operator who has actually engineered rain himself. Practical experience like this is a great convincer.

Nevertheless, when Trevor reported to me the dramatic results of his first airborne use of this technology in Hawaii in 1994, I was highly skeptical. Prodded by him later, my personal evaluation flights in Malaysia proved what an important breakthrough he had made. Being forced down by unforecast violent weather, which appeared out of nowhere, was chastening and convincing. When it happened repeatedly, all doubt was erased.

Certain basic questions often arise from clients concerning these revolutionary methods. Advent of airborne operations triggered still more questions as our capabilities became enlarged. Typical questions were therefore assembled into a written format, primarily for potential clients, but of value as well to a wider audience.

Trevor's answers to these commonly asked questions are backed by 33 years of operational and developmental experience. Technical matters are the prime focus, but highly complex and profound politico-social problems attend the introduction of this new technology and the control of basic natural forces. That aspect is touched on in this presentation.

George K. C. Wuu
Chairman and Chief Executive Officer
Etheric Rain Engineering Pte. Ltd.

Questions and Answers

(Questions are addressed to Trevor James Constable)

Questions that follow are typical of those asked me in my current assignment as Consultant to Etheric Rain Engineering Pte. Ltd. of Singapore. ERE now controls airborne etheric rain engineering operations (AEREO) for commercial use.

QUESTION: America has recently experienced severe droughts, notably in the past year in Texas, and in 2001 in Florida and Washington State. Why has airborne etheric rain engineering not been used to terminate these droughts?

TJC: First, ERE of Singapore does not actually *solicit* business anywhere in the world. Those in need must come to us. I will explain that in due course. Secondly, ERE of Singapore would probably not undertake commercial rain operations anywhere in the USA, because of prohibitive damage liability exposure. Liability litigation is completely out of hand in America. This is tragic. Florida's drought, for example, is within a climatic regime where results with AEREO are virtually sure-fire. Getting results there is not the problem. The real problem is that if ERE were to wipe out a drought that is costing millions daily, American lawyers would not be content with that sweeping public benefit. Essentially bogus lawsuits would be contrived and filed, perhaps many of them, claiming that the company had "hurt" the filers. Lists of alleged damage would be compiled for the courts to wade through. The decent Singapore gentlemen who have backed AEREO would be shunted on to a typical legal treadmill, their good works paralyzed by specious legal action. The litigious character of our culture ensures this will occur, unless a state or federal government would assume all liability. That is not going to happen.

QUESTION: You said that ERE of Singapore does not actually solicit business anywhere in the world. Why is that? How will you become established?

TJC: If business of this all-new character is solicited, a client immediately demands a "demonstration." This is a normal business reaction in dealing with conventional processes and principles. AEREO is not just radically new, but something much more. AEREO excites emotional opposition because of its unorthodox rooting. If confidentiality could be fully protected by a government, there would be little difficulty. In the real world, security evaporates rapidly. Minor government officials and bureaucrats promptly leak the project to their cronies in the media. An opposing firestorm then erupts. Scientists, out-of-power politicians, academicians, pundits, media clones and entrepreneurs profiting from the drought, unite to denounce – in advance – our rain engineering operations.

With the certainty of sunrise, the media can be relied on to act abominably, and to poison the regional public with lies, distortions, ridicule and apocryphal opinions from scientists who still think that there is no ether. We are abused with no fair forum for response. We anchor everything to results. Yet bigoted, ignorant editors deliberately instruct their reporters to trash us. This has been our real world experience. That is why we now require that any government or ruler invite our presence. This changes the socio-political climate completely.

QUESTION: Can such an invitational approach ever lead to a contract?

TJC: Not so far. Disaster conditions have not yet become sufficiently severe, anywhere we are known, for a government or ruler to just ask that we come and discuss tackling their problems. People who have government clout, or the ears of potentates, are likely to hew to conventional views, which "include the ether out," to paraphrase Sam Goldwyn. Someday it will happen and that political leader will become a world historical figure for pioneering the commercial use of airborne etheric rain engineering operations. That will be truly a "giant step for mankind."

QUESTION: Why is ERE reluctant to provide "demonstrations" of AEREO?

TJC: The politico-social aspect of demonstrations is damaging to harmonious working and sound relations with governments and leaders. Nobody wants to be a newspaperman's doormat. Much more daunting however, is the technical aspect of demonstrations. Real danger exists that a demonstration will destroy your own market for AEREO. This stems from the etheric nature of equatorial droughts.

QUESTION: How do you explain that, to a person new to such concepts as ether?

TJC: The atmosphere normally, in equatorial regions like Indonesia, Malaysia, and the Philippines, is strongly cyclic, or oscillatory. Daily passage of the ether in and out of the earth, characterized by the great Goethe as the earth's "breathing," is balanced and healthy. From the Cosmos the earth thus gets the etheric pulse of life, and all is harmonious. Rains come in season. Then comes industrial development, typically as in Malaysia. Thousands of trees vital to the hydrological cycle are cut down. Vast acreages of foliage and jungle are flattened. Asphalt is extensively laid. Factories appear in huge industrial parks. All manner of human commerce that is based upon etherically expansive activity, disrupts primordial harmonies and rhythms. What can now readily happen is that the flowing oscillations that are essential to seasonal rains will be shunted around such areas of lowered etheric potential, which will drain into the higher potential, normal etheric flows, diverting them. A self-reinforcing, self-sustaining drought is thereby created. This is like stopping the pendulum of a grandfather clock, interrupting its normal functioning.

QUESTION: How does this relate to demonstrating rain engineering in such areas?

TJC: An aeroplane or helicopter fitted with a P-gun is able to administer a wide-ranging shock to the regional etheric continuum. This is something like stretching a vast, ultra-subtle guitar string, which is then released. The gossamer elastic etheric continuum is everywhere, penetrating everything. If this stretch-and-release technique is correctly executed, the process can "unlatch" the impeded etheric continuum around the droughted area. Normal etheric cycling tends to resume. Copious rain can ensue. Once the process is understood, and you have seen it happen in the real world, you become wary of "demonstrating" something that can possibly terminate a drought. A successful demonstration could also terminate any possibility of your getting a contract! Stimulated etheric flows can run on long after all stimuli are removed. ERE's corporate video on AEREO describes and demonstrates the process in a far more comprehensible way than any set-piece demo ever could. Clients should have the simple gumption to sign a contract on the strength of that, combined with the decades-long, objective history of this new, scientifically-based art. The first leader who signs up will become a world hero.

QUESTION: Does the engineering process succeed every time, and if not, why not?

TJC: No. Outright, mechanical repeatability is not in the format up to this time. Several short flights may be necessary to establish the correct thrust direction for the required etheric recoil effect to manifest. A delay also usually follows a correct thrust, something common to the stimulation of biological systems. We call this the I.B.D., or Inherent Bioenergetic Delay. In the case of a pinprick to your finger, the I.B.D. is very short. In a stimulus to the etheric continuum, the I.B.D. may be 12 to 18 hours, depending on many factors not yet well understood. This kind of continuum "lag" – with which I became abundantly familiar in shipboard operations – is another reason to avoid mechanical-type demonstrations. In such formats, everyone waits with baited breath for it to rain as soon as the aircraft takes off, and there is an emotional letdown when it does not happen.

QUESTION: Does a prompt response ever happen?

TJC: Yes, it does, and this would seduce the unwary and inexperienced to anticipate such action every time. In Hawaii, I once saw the pilot of our aeroplane take off directly to the east with a P-gun, and trigger off a local shower about a minute after he got airborne. He was right "on tune." That is unusual, however, and the I.B.D. has its way in a later response to a stimulatory thrust.

QUESTION: Approximately how long are these rain engineering flights?

TJC: Usually less than 20 minutes, once on the crucial heading, and very often much less. A prolonged flight does not get you more rain – quite the contrary. The tendency of the aeroplane P-gun is to lose or "slip" its connection to the continuum if the aircraft goes too far, or too fast. The correct process is like operating a very subtle catapult. The energy build-up from the stretch-and-compression of the ether attracts physical moisture. When the "catapult" lets go, rain ensues – often prolonged.

QUESTION: What seems to be an ideal speed at this stage of the technology?

TJC: Substantially under 100 mph, which makes a stable, slow-flying aeroplane right for the job. Better yet is a helicopter, the speed of which can be eased up gradually from the hover. Using a helicopter in Malaysia, results seemed to maximize around 50-60 mph, which is easy on both the helo and the pilot. I have experimented with a biplane capable of over 200 mph, in Hawaii, and this proved that such higher speeds are not effective in this work at this stage.

QUESTION: You are confining commercial AEREO at present to equatorial countries. Why is this, and why do you not extend your commercial availability to the temperate zones of the world?

TJC: The reasons for this present confinement to equatoria are quite clear-cut. The concentration or density of etheric energy in equatorial regions is much greater than in the temperate zones of the world. The rapidity of response, and the magnitude of effects are also greater. The certainty of results is far higher, and virtually surefire if one does not have to produce mechanical, instant, push-button results. Rain is far easier to induce in equatoria than elsewhere, and virtually all life activity in equatoria is more intense than elsewhere. So, until AEREO becomes established, accepted and used by governments, ERE will stay where the risks of operational failure are minimal, and the probability of success very high.

QUESTION: Does this preclude ever using AEREO in temperate zones of the world?

TJC: Not at all. Provings of the basic technology in the maritime mobile format are comprehensive. From shipboard use along the Pacific coast of the U.S, as well as on the high seas, there is no doubt whatever that the techniques are effective in temperate regions. Miles of time-lapse video confirm this. Limited resources however, have not permitted us to explore actual airborne operations in the temperate zones. Thus, we lack the same certainty of results that tropical airborne operations have provided for us. Until that research is accomplished, we will confine ourselves to tropical work. There are plenty of droughts equatorially, worldwide, and always have been.

QUESTION: What significant differences are there between temperate and tropical zone rain engineering operations?

TJC: The most important difference is that the main etheric flows in equatoria run from east to west. In the temperate zones, the main etheric flow action is from west to east. A north to south terrestrial flow of ether also runs in the northern hemisphere winter, and this flow reverses in the summer. There are large differences in temperate zone etheric density and response times, *vis-à-vis* equatoria. Some exploratory etheric rain engineering flight operations, similar to those conducted in Hawaii and SE Asia, are required in the temperate zones as precursors to any commercial work. This will be done someday soon. Certain operational principles valid in mobile maritime work in the temperate zones may need modification in the airborne mode. Some important verifications are necessary, so to speak, before making ERE available to temperate zone contracts. There are people in high places that do not want this technology successful. We cannot afford to overlook anything that could be used to discredit engineering work utilizing etheric force. The principles involved are far-reaching.

QUESTION: Why would any "people in high places" not want such technologies as AEREO to succeed?

TJC: All learned realists, especially those whose real world experience is buttressed by esoteric knowledge, know of the earth's oligarchic powers. Their continued existence and wealth depend upon *control* in key areas. The oligarchs are probably a couple of tiers above mere officials like the president of the United States. Their satellite scientists in the defense establishment secretly know – through Tesla's discoveries – that there is such a thing as etheric force. They also know that through etheric force, technologically applied, more energy and power to do the world's work is available than Man can ever use. This power is eternally available and inexhaustible. Nikola Tesla showed how it could be transduced into human service, and freely broadcast all over the world without wires. Tesla was not permitted to give this liberating power to mankind, as he wished. Blowing up his New York laboratory when they thought he was in it, was the least of many outrages engineered by the oligarchs against this prolific, generous genius. With Tesla sequestered, the oligarchs went ahead and saddled civilization with wired electricity, putting themselves financially astride every kilowatt of it ever consumed. Except that now, the end of this stupendous, but irrational racket is in sight.

QUESTION: How do you see that?

TJC: The cost of extending wired power and its gross, cumbrous technology to the entire world, would require more capital than exists on earth. This is stimulating an urgent search for an alternative way, outside the control of the oligarchs. The only worldly threat to these powerful men is the technological use of etheric force. Such a development would wrest from them their control of the world's energy, politicians and people. Etheric technology transcends and nullifies selfish, greedy and stupid travesties of common sense such as wired electricity, and the countless millions of gasoline engines that now dangerously burden the biosphere. Etheric technology in basic motive power is coming, oligarchs notwithstanding, and etheric rain engineering is only a little zephyr amid the rising winds of change.

QUESTION: Is AEREO a sort of beginning for etheric technology?

TJC: AEREO makes absolutely clear, in the hard world of objective, physical results, that there is an ether, and that the ether can be accessed technically through the correct geometry. There is no technical device on this earth that is simpler than a P-gun. A man with no education – me – has found this out. A Manhattan Project, focusing the efforts of enlightened scientists to the goal of etherically-derived power, would rapidly develop the technology to supplant wired electric power, petroleum fuel, and all the pollutant consequences – environmental, political and financial. That's the wider AEREO scene.

QUESTION: How do you know that additional etheric currents flow seasonally north to south, and south to north as you have outlined? Is this just theory?

TJC: At one time it was theory. You will find the basics of planet earth's ether economy laid out in a classic book dating from 1927. That is Dr. Gunther Wachsmueth's *Etheric Formative Forces in the Cosmos, Earth and Man*. The relevant diagramatic presentations are there, based upon the exact clairvoyant perceptions of Dr. Rudolf Steiner. The formal tendency is to dismiss all of this as occult gobble-de-gook, a view that until fairly recently was partly justified. Once Dr. Wilhelm Reich devised physical equipment with which we could enter the etheric continuum, however, and subject it to appropriate manipulation, we could objectify the resultant effects. These objective effects conclusively proved the existence of these flows.

QUESTION: How was this objectification achieved?

TJC: Simple resonant tubes on shipboard could be used to dam up, locally, the flow of etheric force coming out of the north toward the south in the winter months of the northern hemisphere. Damming up the flow causes a local rise in etheric potential at the damming point, and atmospheric water vapor rushes to that point as though drawn by magnetic action. Now, if there were no ether flow as Dr. Steiner described, absolutely nothing would happen in response to moving empty tubes around. What actually happened, objectively and repeatably, was that in fair weather, a rain nimbus could actually be made to appear on the horizon right on the magnetic north vector. This accretion could be made to appear from nothing, fairly in the center of a video frame aligned on magnetic north. By maintaining the alignment, the accretion could be further densified until regional rain appeared, "out of nothing." This rain mass would come right over the ship, drenching the porthole glass. The visible scenario videotaped in front of the ship was totally transformed in under an hour. The entire process was recorded on time-lapse video, and could be reviewed a hundred times over if desired. Such sequences were engineered many times, and made the existence of this seasonal flow utterly irrefutable. The same apparatus and procedures in the summer months verified unerringly that the summer flow came out of the south and could be similarly interdicted and "brewed up" – the term we have used for this type of operation. Dr. Steiner's etheric blueprint was thus verified in the real world a long time ago, in its essentials, and undergirded literally thousands of mobile maritime rain operations. By the grace of God, I was a professional aboard a large, modern and fast ship on the high seas – working in pristine conditions. I was therefore able to make these objective findings. In turn, and when I devised suitable airborne equipment, all the fundamentals proved valid in the airborne mode – only much more powerful that way, in their application and results.

QUESTION: What does it cost a client to engage the services of E.R.E., and how are contracts usually handled?

TJC: Each client is in a particular place in the world. That place has its own special character in terms of climate and weather. The client's difficulties tend also to be specific to him. Each client is differently placed economically, and most political leaders initiating rain engineering contracts have to contend with their political opposition, who will usually attack the leader's doings without a care for the public's sufferings. The price charged to the client in each contract normally has to factor in all these elements, and others, to reach a feasible price. Mr. George Wuu, ERE's chairman and chief executive officer, normally does not consider any contract under one million dollars. In an age where governments talk in trillions, that is a small amount of money for big benefits. Regional drought can easily cost $10 million a *day*.

QUESTION: What about the handling and contract terms. Can you give a general idea of those factors?

TJC: Standard contracts with ERE are very straightforward and minimize risk to the client. A territory or "target zone" is specified, and the price of the rain and the contract term are agreed on. A fairly standard example would be $10,000 per millimeter of rain. That seems exorbitant, if you have ever seen one or two millimeters of rain in a gauge. You must remember, however, that we are dealing in this type of contract with a *territory*, perhaps 100 square miles in extent. One mm of rain over such a terrain is more than 69 million gallons, or 214 acre-feet. That is a very reasonable price for water in a droughted environment, but we usually engineer rains of more than 1 millimeter.

QUESTION: How is payment effected under ERE's standard contract?

TJC: The client must deposit with a solid financial institution, and preferably with a recognized international bank, the full amount of the contract, say, one million dollars. This deposit or credit is escrowed, with the conditions of its disbursement to ERE written into the contract. Inspection of rain gauge amounts, and release of any funds accordingly due to ERE, is under the control of the Societe Generale de Surveillance, a long-established and internationally respected Swiss organization. Among their wide range of services to business and governments, SGS includes environmental monitoring. Under this rational system, we must be paid if we deliver rain as per contract. We are not paid if we do not deliver, and if we do not deliver within the contracted time, SGS will release the escrowed funds back to the client. Thus the client cannot lose. Similarly, we will not be in the position of engineering rain successfully, whereupon the client decides – in the absence of such a binding escrow – that he does not need to pay us now that he has the rain. This has happened, but is precluded under the arrangements described.

QUESTION: Isn't there some risk in tackling rain engineering projects without some trial period or tests in a particular region?

TJC: The standard contract includes a mobilization clause, carrying an upfront fee. Obviously, ERE cannot absorb mobilization costs, sometimes involving extensive air travel and other transport costs, and making this also contingent upon rain results. The mobilization charges also defray any exploratory small operations necessary to establish the feasibility of the contract's main objective. Etheric rain engineering is not like turning on a tap, or using a hose. Each region likely to experience drought will have its own character and quirks. Such small mobilization and exploratory operations are likely to be reassuring to the client if they prove fruitful. Mental hazards connected with the main contract purpose and term, and its novelty, can be dispersed this way. With the information I have given here about contracts and agreements, there can be no doubt that ERE policy is to protect its clients and their interests, and also the ordinary citizens whose lives are being ruined by drought. They are the ones actually funding the ERE contract.

QUESTION: Does ERE require that the client assume damage liability in its territory?

TJC: Yes, that is in the contract. This is not a sticking point or even a matter of discussion with governments, other than in the USA. That there may be damage is simply accepted by most governments as collateral to the regional benefit anticipated. An omelet cannot be made without breaking eggs. Furthermore, outside the USA, drainage and flood control facilities are usually far less well developed than in the ordinary American community. The result is that hazards and damage are likely to be much greater, working in less-developed areas of the world. This overhangs everything you are seeking to do in developing regions and countries.

QUESTION: Does the vulnerability of ordinary people to flooding in developing countries bother you personally?

TJC: My concern about that is unceasing. Indeed, it justifies having some sort of small, preliminary trial operations prior to the full contract effort. One may find local and regional conditions intractable and not amenable to etheric rain engineering. In that event, you would not pursue a full contract. So also, it may turn out that etheric conditions are labile and easily unbalanced to cause extended rainfall. Great care would be vitally necessary. The security and welfare of the civil populace are always major concerns. Knowing at the same time that the drought is killing those poor people and their children does not lessen the pressure on a rain engineer. Little of this is comprehended by our critics.

QUESTION: Rain engineering and smog control are obviously different applications. Can you describe the actual differences in these two kinds of etheric engineering operations?

TJC: Rain engineering has developed into a highly mobile, airborne operation – exclusively devoted to engineering rain. Smog control remains essentially a non-rain operation from fixed bases, in or near a smoggy region. Different techniques and different apparatus are used in each case, and the goals are different. The common functioning principle is accessing technically the ether. All of this is unorthodox and unaccepted by formal sources and official authorities.

QUESTION: Can you provide some detail on this, based on actual operations you have carried out successfully?

TJC: The initial development of etheric rain engineering was due to the visionary genius of Dr. Wilhelm Reich. Everything proceeded from his inspired invention of the cloudbuster. This device is an array of metal pipes mounted on a turntable platform. This permits the pipes to be placed on any bearing at any desired elevation, and the array can be moved while aimed into the sky – just like an Oerlikon gun battery. One end of the pipes is grounded into water, usually via a bundle of BX cables. Everything involving the engineering of etheric force down to this day – including my current single resonant tube device – is descended from Dr. Reich's cloudbuster invention. He is the grandfather of it all, and one of the greatest men of all time.

QUESTION: Is the cloudbuster still used today?

TJC: Worldwide, many experimenters use the cloudbuster, including some scientists. Such operations in northeast Africa, for example, were highly successful under Dr. James DeMeo, creating vast, regionally beneficial lakes. My work proceeded in a different format, due to my having the unique opportunity to work experimentally aboard ocean-going ships, from 1967 until 1992. This is a radically different activity from fixed-base work ashore. Many developments and insights came about because of the mobility factor and shipboard operations under pristine, high seas conditions.

QUESTION: Did you ever use cloudbusters in combating smog?

TJC: Not deliberately, no. And not in operations pre-filed with the U.S. Government. In my early, shore-based work, using huge arrays of water-grounded pipes and Dr. Reich's basic findings, my team was able many times to clean up the filthy atmosphere of the Los Angeles Basin region. This was achieved adjunctively to rain engineering objectives. Logistical and geographic problems in those days made sustained operations exclusively for smog clearance purposes intolerably burdensome. Later developments of biogeometric equipment, a legacy of extensive maritime mobile work, were especially suited to smog clearance operations and allowed us to give attention to such activities.

QUESTION: Does this not bring us to the differences between smog clearance operations and rain engineering operations that you started to describe?

TJC: Yes. In order to maintain shipboard experimental work, I was required by shipboard discipline to quit using water grounding, something fundamental to Dr. Reich's original designs. Water from my equipment, cascading down from the flying bridge of SS Maui at all hours, became an intolerable nuisance, and a threat to the privileges granted me. I had been allowed to turn the flying bridge into a traveling laboratory. This was no small privilege, on a huge commercial vessel. I was desperately reluctant to leave the seeming security of Dr. Reich's pioneer formatting of this work, but I had no option. When water grounding was abandoned, our thinking was forced into new, original channels. This revolutionized the art and technique of engineering etheric force in weather work.

QUESTION: How was it possible to operate without any water grounding?

TJC: Geometric forms and structures have a long history of involvement with etheric force, the pyramids being a major example. Chinese philosophy and other cultures abound with such connections. As Westerners, we knew that cones, like pyramids, focused etheric force and emitted that force in a coherent beam from the cone apex. Sensitive people can directly feel this kind of emission. My highly valued associate, Lou Matta, then serving as Chief Engineer of SS Maui, is a serious student of Tal Chi and also of sacred geometry. We made the cones with our own hands, and mounted and juxtaposed them in various ways. We built devices that emitted beams of etheric force and thus did not need a water ground. We added the component of rotation, via small electric motors. This proved a winning combination when added to the 23-knot velocity of the ship. The most effective device we used for combating smog – the Spider – grew out of this work.

QUESTION: Can you describe a Spider in general terms?

TJC: There are numerous photographs of Spider units in *Loom of the Future*, a book which overviews my weather engineering work. [See also page 191.] There are also many versions of the Spider seen in the publicly-released video, *Etheric Weather Engineering On The High Seas*, wherein they are actually seen working at sea, engineering rain despite high barometric pressure. The Spider is essentially geometrically-correct cones mounted so that they project – into the sky – etheric beams from their apices. These cones are rotated with a small motor. With this arrangement, you generate vortexial motion in the ether.

QUESTION: In a smog clearance operation, you would have to operate a Spider from a shore side fixed base, wouldn't you?

TJC: Yes. But you must remember that the ether itself is not stationary, but in constant motion. The old idea of some famous scientists' a century ago that the ether is stationary – before they threw out the ether entirely – is false and misleading. Therefore, when you operate a Spider from a shore base, it is stirring vortices into an ether that constantly moves through the Spider site and all around it. In this way, *vortex strings* are steadily emitted from a functioning Spider, and these vortex strings travel on over the earth indefinitely – carried "downstream" from the Spider site. Sometimes they cover hundreds of miles. That is why you have to place your Spiders strategically to clear smog out of a targeted area. You want vortex strings to cross the target area, carried on one of the etheric flow vectors prevailing regionally. To cleanse Los Angeles, therefore, you need to have such Spiders positioned west of the worst smog areas, and in summer you also need to have Spider stations south of the L.A. Basin. This will ensure vortex strings crossing the target area from south to north.

QUESTION: How do you theorize that vortices in an alleged, superfine, unseen "ether" can reduce or combat smog?

TJC: Etheric vortices are of two kinds: implosive and explosive. Conventional technology depends upon explosive forces. This is not only the trillions of explosions of internal combustion engines, but all the resultant, etherically expansive activity essential to electric and electromagnetic technology. The explosive activity is broadband, massive, ever-multiplying and hostile to life and health. Along with the directly polluting particulate matter, the sun exacerbates all this. There is nothing "ordered" about this etherically expansive chaos. It is hot, dry and horrible. Metro smog is one consequence.

QUESTION: Then what can Spiders do about such a situation?

TJC: Spiders emit endless strings of *implosive* vortices, which are cooling and contractive. They are also harmoniously *ordered* activity, by virtue of the geometry of the cones. That geometry is based on Golden Section ratios, which manifest in everything living. These vortex strings appear to have the same properties in miniature as the great implosive vortices we call typhoons and hurricanes: they entrain material substance and drive it to the point of the vortex. We believe – with strong objective backing – that engineered implosive vortices entrain particulate matter that has been boosted into the lower atmosphere by all the explosive activity of civilization, and drive that material substance back to the ground. The clarity of the air in Los Angeles when 14 Spiders are operating is wonderful to behold.

QUESTION: What about temperature, in a regional smog clearance operation?

TJC: Experience has proved, in operations in L.A. in 1987, 1989 and 1990, that temperatures come down considerably. Smog is drastically reduced, and the delightful quality of the weather during these Spider operations evokes rhapsodic comment from the media. They think these paradise conditions are acts of God, or blind luck. They would condemn out of hand any suggestion that conditions were being engineered. We have documented our successes not alone with official statistics from the Air Quality Management District, but with contemporary media clips recording dumbfoundment at the unbelievable weather and low smog. In the month of July 1987, for our pioneer smog clearance effort with Spiders in southern California, we had only three primitive units, but they exerted staggering influence on regional conditions, and documented themselves, big time.

QUESTION: Can you explain what you mean by "documented themselves?"

TJC: The California Air Resources Board set up an expensive project that same 1987 July, to investigate smog. The venture cost millions. They imported a host of "smog scientists" from all over the USA and abroad to study smog. Special monitoring bases were manned with these people at more than a dozen sites in southern California. Trouble was, those primitive Spiders were dramatically effective. Smog was so reduced that there wasn't enough of it for any of their studies to have statistical value. Consequently, the project was *aborted* and the scientists were sent home. Nothing like this humiliating debacle had occurred in the history of Los Angeles smog, and the TV cuts we have preserved that describe this fiasco are worth their weight in laughs.

QUESTION: Did you ever hear from the authorities about your activities?

TJC: Never. Never have they deigned to give an inch on such matters. The Air Resources Board and the Air Quality Management District were always evasive in every respect, despite being advised in advance, every time, of our presence in the smog equation, when we filed our pre-operational advices with NOAA.

QUESTION: You've given a word picture here of a smog clearance operation. Can you now describe the contrasting scenario around airborne etheric rain engineering?

TJC: Let's assume a roughly counterpart setting to the smog clearance activity: a large metropolitan region in Southeast Asia, suffering from severe drought and crop damage. Water rationing and demoralization harass the metropolitan population.

We are going to use a single, resonant tube of special design and sensitive construction called a "P-gun." This device utilizes no chemicals, no electronics, batteries or electromagnetic radiation in any form. We place this simple bio-geometric device aboard a helicopter – the ideal aircraft for this task – take off, and head toward magnetic east at about 50-60 knots. There are immediate consequences from the intelligent entry of this device into the etheric continuum. Any cloud banks visible may change rapidly in appearance and density, some may spring little "leaks" while you watch. All around you is evidence of what we call "insertion shock" – the physically observable impacts and impressions of etheric response and disturbance being translated down into the atmosphere around you. What is required is to press steadily eastward, and gently vary the heading of the helicopter until the "tuning" of your helicopter/P-gun combo interdicts the regional etheric flow coming at you from the east. This etheric flow, of course, is invisible, so you only know that you have "snared" it by what happens objectively around you. Thousands of maritime mobile rain operations have taught me what to look for, and in the airborne mode you really have to keep your wits about you because of the rapidity with which violent weather can come out of fair sky conditions – with no such activity forecast.

QUESTION: How long do you have to fly in the helicopter or light plane to instigate rain?

TJC: Usually less than 20 minutes "on vector." In equatoria, short flights on key vectors, as described, are the most effective. If correctly handled, they will push the ether flow back, causing an immense buildup of etheric potential behind this backed up flow. Atmospheric moisture accretes heavily to this etheric build up, and you see this as rain-bearing formations. You should by then have deactivated your P-gun and be racing rapidly back to base. You may not, in such operations, see any immediate accretions of rain-bearing cloud, but your experience may tell you that you have been successful in pushing the ether a *long* way back: like an ultra-subtle compression effect. Rain from such a flight may not come in until hours later, and will probably last for several hours when it does come. I can say on the basis of practical experience that these airborne etheric rain engineering operations present abundant proof of the theories of etheric functioning advanced by Nikola Tesla, Dr. T. Henry Moray, Viktor Schauberger and others who have tapped, practically, the stupendous power latent in the ether. Once you get the ether triggered off in an airborne rain engineering format, it's likely to just keep a-coming. Tesla knew about this etheric property.

QUESTION: Is there a certain amount of ART in etheric rain engineering?

TJC: Of course there is. The acumen required to do it successfully is an artistic faculty. In my case that faculty was developed by thirty years of practice – just like the talent of a musician is magnified by study and practice. In the future when we know more, we will be ourselves *changed by our knowledge of the etheric*. That will change our approach, our outlook, our capacities, our technology, our attitudes and our world conception. Today's world will be looked back on as a junk civilization.

QUESTION: Is there anything else you can say about airborne etheric rain engineering?

TJC: I could say a lot, but what I have given here should suffice to establish the profound difference between rain engineering and smog clearance. ERE of Singapore does not any longer pursue smog clearance contracts, but would probably "throw in" the technology with any government that entered a long term rain contract with ERE. I think that is possible.

QUESTION: Does ERE of Singapore actually prefer long-term work to drought breaking?

TJC: Definitely so. The ideal purpose of etheric rain engineering is to ensure that droughts never get started and never become catastrophes. The gentle and intelligent use of airborne etheric rain engineering would keep the whole etheric scenario harmonious and life-positive. Most of today's political leaders allow droughts to ravage their lands and people before taking panic-driven, desperate measures to combat drought. Prophylaxis is the most effective way, the cheapest way, and the most beneficial and constructive way to deal with drought.

QUESTION: Do such "prophylactic" operations against drought usually involve long-term work?

TJC: Yes, but I do not see this as continuous engagement. Rather would it likely be occasional operations to offset incipient drought, somewhat in the fashion of a fire brigade. Nevertheless, my long involvement with this new engineering art has convinced me that its greatest and inarguable impact occurs in protracted operations. The statistical base for comparing one season or year with another is much broader than attempting to make a convincing case in, say, a given week. Our greatest single victory was the 1990's Operation Clincher, where we operated 6 months – the entire smog season – against the statistical records of all prior seasons, and, in the entire four-county Air Quality Management District. Bringing smog down by 24 percent across the board, for the whole season, was a major triumph unobtainable in any other way. More to the point, the operational target of 20 percent reduction was filed with NOAA in advance.

QUESTION: Did you advise the smog authorities of your project and its goals?

TJC: We always did that. The Clincher filing had them laughing in the meteorology department, but the last laugh was mine, when seasonal statistics were released. Seasonal smog had been devastated. The lesson is that extended projects can stand out statistically, whereas short, drought-breaking operations probably will not. Such successes can be verbalized away by evasive officials as something that "would have happened anyway." Only by such dishonest obfuscation can your documented, pre-filed engineering goal be deprecated and masked.

QUESTION: Has there ever been any commercial approach made to ERE for long-term application of this technology to an otherwise unachievable goal?

TJC: The most intriguing proposal for this work I have yet seen, was broached semi-officially from southern Baja California. This is a bone-dry, scorching peninsula off Mexico's west coast, in about the same latitude as Oahu, Hawaii. The general idea was operations over a one-year period, to gently initiate the restoration of agricultural areas ruined by chronic drought. With sea on both sides of this peninsula, southern Baja is an ideal place for a comprehensive, objective and irrefutable proving of AEREO. I could readily envision, through such an extended and rational program, a gradual greening of the whole area, which is now parched, barren and sterile. The menaced subterranean water supplies, now being infiltrated with seawater, could be renewed. Verdancy and productivity could be steadily restored. A magnificent venture like this makes mere drought-breaking seem almost juvenile. I pray for such an engagement to crown my life.

QUESTION: What happened to all this?

TJC: The project is in limbo as of May 2001, but may be reactivated soon. Sometime, somewhere, a visionary leader facing such a crisis will "roll the dice" on this environmentally pure technology. He will thereby bless his people and the planet as well. That leader will also achieve world historical status for himself. Orthodoxy comes out against us, but the incontestable truth is that with drought, Mother Nature laughs at conventional "high tech." She laughs also at those who are in thralldom to this clay-footed modern god. Mother Nature seems to prefer someone like me – a guy with no education – who will listen eagerly to her subtle, living whispers.

Chapter 14

RADAR IN ETHERIC RAIN ENGINEERING DEVELOPMENT
An Autobiographical Account

Radar played an indispensable role in developing this new scientific art, especially after 1978. During the late 1960s through the mid-1970s, fixed-base regional rain engineering projects were guided by satellite visual loops, radar loops and radar facsimile maps available from US Government sources. In this period, weather engineering operations utilized batteries of over 100 large-bore, water-grounded tubes, emplaced mainly in remote desert sites in southern California. The effects wrought by such installations were sometimes so extensive geographically as to strain credulity. The operations overall were ponderous and difficult to manage.

Operational effects were often readily discernible in visual satellite loops, and sometimes on government radar. The sophistication and coverage of weather satellite technology continually progressed. Documentation of operations was thereby available in the 1970s, including comprehensive weather radar that did not exist when the late Dr. Wilhelm Reich pioneered weather control technology decades previously. That Dr. Reich was able to develop this technology as far as he did, without the abundant documenting aids available today, generated a profound, permanent respect for the genius-level awareness of this great scientist.

Operating on shipboard in the late 60s and on into the 70s as a Radio Electronics Officer, vessel radar was among my professional technical responsibilities. This was a radical change from dependence on faraway government radar facilities. I was in a completely different environment in every respect. Gone was the highly resistant, virtually etherless aridity of the southern California desert and the Los Angeles region. Gone also were the costly, long road trips to tend and adjust the huge, water-grounded batteries of tubes then in vogue. Not only were the pristine ocean expanses regions of highly active etheric forces, they could be scanned continually with what was, in effect, my own radar. U.S. Merchant Marine ships normally run their radars

24 hours a day while at sea, so there was no waiting for facsimile schedules or satellite information. Engineered effects could be studied as they occurred. Excluded from the new set-up were the immense, water-powered batteries of weather guns with which I had begun my weather engineering investigations in 1967-8. Drastic downsizing of equipment was mandatory, at the same time as radar came to my personal hand. I would have to learn to make do with a single tube, on shipboard.

The new format in which I would henceforth operate, anchored to and dependent upon radar evidence, suddenly became local. This was an almost revolutionary change. The massive installations used ashore could, and often did, produce mind-boggling long-distance effects. Even associates most sympathetic to the projects had difficulty accepting results that sometimes influenced hundreds of square miles. In 1977 for example, a directional change of our equipment (150 guns) located at Thousand Palms Oasis in the desert interior of California, produced a sudden, 180-degree wind change in Santa Barbara, on the Pacific coast 150 miles distant. These dramatic effects occurred fortuitously, and not on purpose, when the city was about to be consumed by the ferocious Sycamore Canyon fire. The wind-driven holocaust was stopped in its tracks and blown back up the hill by a cool wind off the Pacific Ocean, on the same vector as our changed operational direction from the desert. This operation is well documented. Furthermore, a scientist from the University of Southern California who was a neighbor of mine in San Pedro, happened to be offshore from Santa Barbara in the research ship Vallero IV. He saw the overpowering of the fire, and was duly impressed. “Big John” usually joshed me unmercifully, but not on this occasion. Many other friends had difficulty accepting that these remote effects were possible with such crude, almost ridiculously primitive equipment. This was a conceptual problem of some magnitude.

The maritime rain engineering situation left all this behind. What people could not grasp or comprehend because of its immense scale could now be brought down to horizon distance, and within reach of ship’s radar. It was also recordable with cameras and the portable video recorders then coming into use. Operating at horizon distance held the possibility of eventually convincing even the skeptical and faint-hearted that etheric rain engineering was indeed an objective reality, as Dr. Reich had shown decades previously to an impervious world.

Mankind was technologically tapping the ether of space. Operations like the Sycamore Canyon Fire, and Kooler in Los Angeles, showed that this was a continuum of high power, able to influence millions of tons of atmosphere. The philosophical situation had also changed in favor of proving the ether's presence as a physical entity, and its accessibility via geometry. Proving the ether mathematically, or disproving it mathematically, with arguments back and forth, had been going on for a century. All of that now seemed irrelevant. Radar proved that a person knowledgeable in basic etheric functions could physically "get hold" of the ether. Radar provided the evidential and methodological key, linking the rain that was being engineered to what could be electronically and visually demonstrated. Denying what was objectively happening was like saying there was no such thing as a punch in the nose. My own radar installation, so to speak, could be consulted at any hour of the day or night.

The pristine, active ocean environment provided abundant rain engineering results, both visual and electronic. Forced reduction on shipboard in the size of the installations used ashore, proved to be the greatest of all blessings. Simplification and size reduction, from that point, have continued down to this day. Physically smaller rain engineering gear produced radar-detectable results as far away as 30-40 nautical miles. The behavior of these rain formations relative to the ship's movement was plotted and studied. Definite evidence appeared of a basic law of etheric potential originally formulated by Dr. Reich: Etheric energy flows *from low* potential *to high.*

Dr. Reich termed the energy he discovered *orgone*. There is no doubt on my part that orgone energy is functionally identical to the chemical ether that is technically accessed during Etheric Rain Engineering operations and makes the entire process work. There are at least forty different names, in various cultures and epochs, for this same force. Dr. Reich termed its basic law the Law of Reversed or Orgonotic Potential. Rain engineering functions become achievable by understanding and applying this law. Radar played an indispensable role in elucidating the working applications of this law, which lie behind all successful etheric rain engineering and weather control. Radar was totally objective, divested of all emotion. All-new facts were disclosed in the glowing phosphors of those marine radar screens. Without this irrefutable testimony, and constant access to such testimony on the empirical pathway, progress could not have been made.

As long as I was being assigned to different ships for every voyage, no consistency of operational activity was feasible. Most of the ships served on during this period were fitted with single, 10-centimeter radars. These particular radar units tended to penetrate rain formations for vessel-to- vessel safety, rather than provide echoes from such rain. Occasional assignments to vessels fitted with newer, 3-centimeter radars convinced me that in such units lay a dependable means to develop maritime mobile etheric rain engineering. Three-centimeter units, when well-tuned, returned echoes from rain formations out to 40 nautical miles, and sometimes more when the rainy systems were large and reached miles up in the atmosphere. The extent, intensity and motion of rain formations were well-defined by 3-centimeter radars. Such technology came into vogue mainly because of its superiority over 10-centimeter units in piloting functions, such as detecting small buoys, through the higher resolution the shorter wavelength provided. My occasional encounters with such 3-centimeter radars during this period of temporary assignments exposed me to alluring glimpses of what was possible – only to be followed by my transfer away from such 3-centimeter units.

When I received permanent assignment to the SS Maui, flagship of the Matson Navigation Company fleet, on her maiden voyage in 1978, all this changed. There would be no more transfers every two or three months. On vacation leave, any material or equipment could be left securely aboard in the vessel's abundant storage spaces. Cooperative understandings and arrangements could be forged with permanently assigned shipmasters and officers. Matson was a wonderful, "people-oriented" company. Furthermore, the Maui had two modern Sperry radars: 3 centimeter and 10 centimeter. The 33,000 hp ship made 23 knots, which was to prove itself an important operational asset. The Maui was also an official station of the World Meteorological Organization, and was equipped with all required and officially approved meteorological instruments. The barometer and approved wet-and-dry bulb thermometers were eventually integrated into the numerous video sequences that were taped aboard.

Commodore C. C. Wright Jr., senior Matson Lines master at that time, was in command of the Maui. Intelligent, open-minded and a consummate gentleman, this former captain of Matson passenger liners readily assented to my carrying on rain engineering experiments aboard the ship. Commodore Wright took a lively interest in results. The radars would provide objective confirmation of any rain formations that were developed. Following is an account of an early Maui rain engineering success, excerpted from *Loom of the Future*, a heavily illustrated interview book dealing with the overall development of etheric weather engineering. From this beginning, aboard the Maui, a new epoch of etheric rain engineering opened, based on radar's trustworthy guidance. Stabilization of etheric rain engineering investigations, by virtue of my permanent assignment to the vessel, completed the new scenario with all its promise. The short excerpt which follows describes an early operation aboard the ship:

> Rain engineering aboard the Maui started with a single, rack-mounted, 4-inch diameter tube, equipped with a water ground. This was absolutely the barebones, rock-bottom type of cloudbuster device. I started out by simply pointing the tube aft – that is, aiming it over the stern along the fore-and-aft line of the ship. This followed Dr. Reich's admonition always to aim the device in the *opposite* direction from which you want the rain to come. The ship was on a southwesterly course for Hawaii. We were in about 1020 millibars at the barometer, and around 70 percent relative humidity, so the likelihood of rain was virtually nil.
>
> The 3 centimeter radar display started out completely clear of rain echoes, until about 40-50 minutes after this simple assembly was aimed and energized. To my happy astonishment, small, sharp squalls then began to materialize about 10-12 miles head of the vessel, right on our course line. These squalls fell away on each side of the ship's course, giving rise to an arrowhead type of formation, which I will sketch. As the rain squall echoes passed down each side of the ship, many miles out, they were obviously losing their cohesion and reflectivity. Rain echoes faded rapidly once the squalls passed abaft the beam on each side of the ship. All the while, more rain squalls continued to generate at the same point dead ahead. The exact distance of this point was readily ascertainable – and was monitored – with the range marker control. The engineered echoes always came up right on this electronically-marked point. A geometric distribution of this engineered activity was taking place before my eyes. Radar does not lie.
>
> When Commodore Wright came on the bridge to write his night orders, he went straight to the radar and took it all in immediately. "I'll be darned," he said, "a blooming arrowhead." He suggested right away that I turn off the

> cloudbuster unit, and see if the arrowhead formation faded out. I did this, and the collapse and dissipation of this engineered, artificial system commenced immediately the water was turned off and the tube disoriented. You could see the whole display lose energy and reflectivity very rapidly. In half an hour, only the faintest remnants of that once vivid arrowhead formation remained abeam, and all generative activity dead ahead of the vessel had ceased.
>
> We then tried reactivating the unit. The formation did indeed return in an hour or so, but in diminished strength and clarity. I would caution here against attempting to transfer to this new engineering art – dealing with an incredibly subtle force – mechanistic procedures pitch-forked over from the regular science and engineering fields. An entirely new attitude is necessary. We are dealing with something that is alive. We need to ensure that reality and technological utility are not extinguished or buried, by compulsively demanding compliance with gross, on-off expectations. The latter are typical of the cruder aspects of intelligence. The need to change ourselves is universally apparent in this decadent world, and weather engineering makes that demand on us. With the ether, we are handling a baby right now. Great care will give us a sturdy infant in due course.

From this beginning, the experiments remained solidly anchored to radar results, right up until the time my sea service ended in 1992. When Commodore Wright retired, his successor was Commodore K.R. Orcutt USNR, a gentleman of formidable mentality, intelligence and technical skills. In the course of my seagoing career, I had sailed with about 60 different shipmasters, under three different flags, in all kinds of vessels ranging from RMS Queen Mary on the North Atlantic, down to Liberty ships and tramp ships. As a result of this experience, I knew a good shipmaster from a mediocre one. Commodore Orcutt was without a peer in my experience. He knew radar, both operationally and technically. He gave quiet support to my work. He ensured it was known that no one was to interfere with my installations on the Maui's flying bridge. There were those aboard the ship, emotionally upset by its presence and effects, who would have clandestinely thrown it all overboard.

From time to time, the Commodore offered significant, quiet suggestions – such as suggesting to me that I get the technology airborne – that were intriguing. He also occasionally asked me to intervene operationally in foggy conditions, which was usually successful, and sometimes spectacularly so. Scientifically-minded and technically skilled in four or five disciplines – including computer programming – Commodore Orcutt's eye was always on those radars. "Radar does not lie," was a phrase he much favored.

FIRST USE OF CLOUDBUSTER ABOARD SS MAUI, 1979

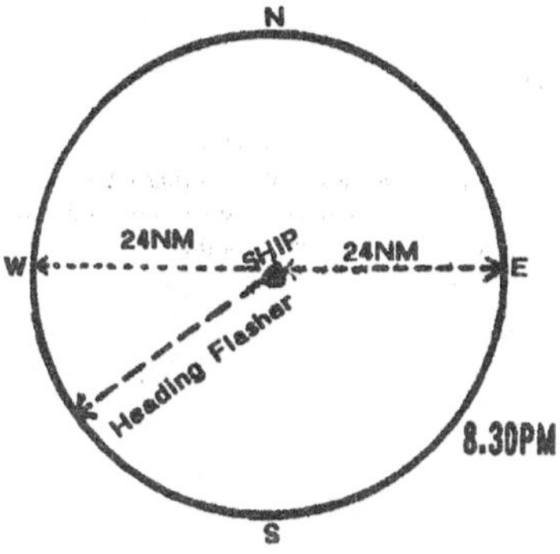

VESSEL IS AT SCREEN CENTER.
SCREEN DIAMETER IS 48 NAUTICAL MILES.
HEADING FLASHER IS DIAGONAL DOTTED LINE, SHOWING SHIP HEADING SOUTHWEST.
SHIP'S SPEED IS 22 KNOTS.
BAROMETER IS 1025MB.
REL.HUMIDITY 75 PERCENT.
RADARSCOPE SHOWS NO RAIN ECHOES AT CLOUDBUSTER TURN-ON, AT 8.30PM.

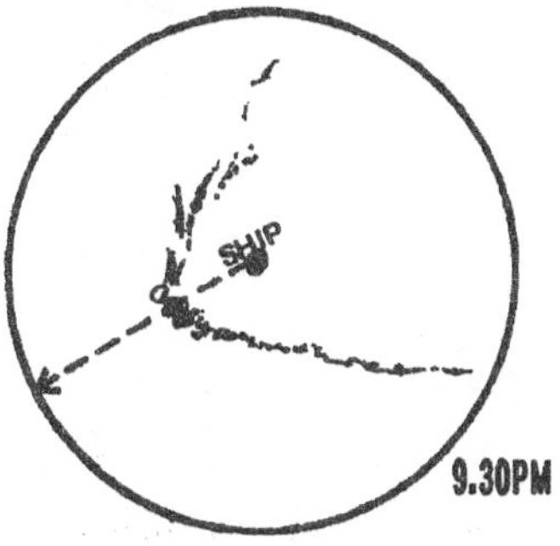

RADARSCOPE AT 9.30PM. SHOWS RAIN ECHOES THAT DEVELOPED AT 10-12 MILES RIGHT ON HEADING FLASHER. THIS WAS IN RESPONSE TO CLOUDBUSTER OPERATION, AIMING OVER THE SHIP'S STERN. RAIN ACTIVITY SPREAD OUT IN ARROWHEAD FORM TO EACH SIDE OF VESSEL, DIMINISHING SHARPLY ONCE ABAFT THE VESSEL'S BEAM. GEOMETRY OF SCENARIO REMAINED UNCHANGED UNTIL CLOUBUSTER SHUT DOWN ON SUGGESTION OF COMMODORE WRIGHT, A WITNESS TO THE ACTION. CLOUDBUSTER SHUT DOWN AT 9.40PM.

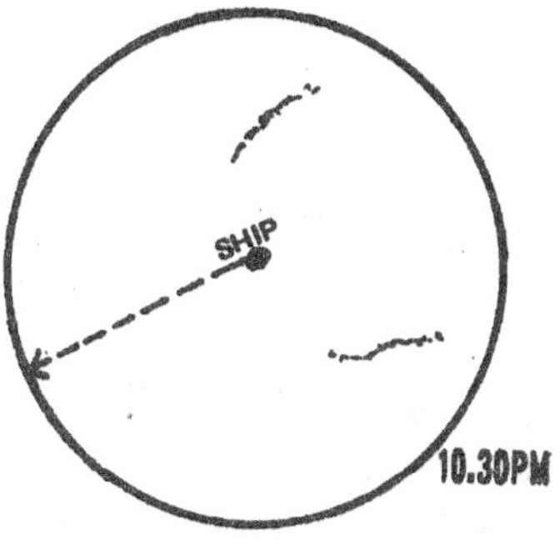

RADARSCOPE AT 10.30PM. RAIN ECHOES NO LONGER GENERATED DEAD AHEAD. ARROWHEAD RAIN FORMATION DISSIPATING, ONLY FRAGMENTS REMAIN, FALLING ASTERN. ENTIRE INCIDENT IS DESCRIBED IN THE TEXT. ENGINEERED RAIN ECHOES AROUND THE MAUI ALWAYS FELL INTO GEOMETRIC PATTERNS, OF WHICH THE ARROWHEAD WAS ONLY ONE.

In the first years aboard the Maui, the rain engineering research continued with the use of two or three large-bore tubes on the flying bridge that were water-grounded. Extensive, irrefutable time-lapse videotapes were made of operational results. Radar guided this work, and confirmed the reality of the etheric laws involved. When proceeding on southwesterly courses outbound to Hawaii at night, with weather guns in function, radar often confirmed the presence of vast rain lines 30-40 miles in extent, perpendicular to the ship's course. Without radar we could not otherwise have seen, or understood these accretions of thousands of tons of water. By aiming the muzzles of these large tubes abaft the beam, with the depressed, water-fed end lying forward in the direction of magnetic west, the etheric flow coming out of the west toward the east, could be locally blocked. The low-to-high flow law of etheric force, and the forward drainage of the ether away from the ship created this strange, "backwash" effect. The resultant local etheric blockage near the surface produced an elevation of the etheric potential over the ocean ahead of the ship. Atmospheric moisture would be drawn as though by a magnetic force to this area of higher potential. Rain materialized in the high potential region, whereupon it became visible electronically on radar. 3 cm radar clearly showed the rain accreting and not moving. Such operations as these would not be attempted unless we were proceeding in rain-negating barometric pressure of 1025 millibars or more. This was all being done at night, when the etheric continuum is most amenable to such manipulations.

Commodore Kenneth R. Orcutt

TJC (standing) with Com. Orcutt

Etheric force from farther west meanwhile, beyond the blockage area, would run toward the high potential blockage, drawing still more atmospheric water vapor and causing the saturated region to spread each side of Maui's course line. From these accretions, deluges would descend to the ocean surface. This whole sequence was unerringly revealed on radar. The convincing climax would come when the Maui, with etheric force still draining forward and away to both sides from the shipboard gun, would sail right through the engineered front in virtually rainless conditions. On moonlit nights when carrying out this kind of operation, we could see rain descending copiously on both sides of us while we sailed between these walls of deluge with only drizzles on the Maui. This aspect of our operations we christened the "Moses Effect," because of the parting of the waters. Countless dozens of deck officers and others saw the "Moses Effect" during their Maui service.

All of this was depicted unequivocally on radar, many times in similar circumstances, pressed down and running over. Dr. Reich's Law of Reversed or Orgonotic Potential was valid, as was the west to east etheric flow in temperate latitudes. A photograph (next page) is included in this section of the Maui's radar screen, taken during one of these operations while en route to Hawaii from Los Angeles. Another photograph in the presentation shows a similar kind of operation near Oahu, Hawaii, but with the opposite, east-west flow of etheric force that prevails in the tropics, blocked by rain engineering equipment on the Maui (page 153). Thus was the ether engineered to "brew up" a massive rain formation. This would contain countless thousands of tons of water, and extend horizontally over 30-40 nautical miles. Once again, it is 3cm radar that allows the entire happening to be analyzed and understood as the application of a law new to mankind. This law may one day apply etheric force to lift a craft into the air as easily as it now lifts thousands of tons of water. "Fuel" is not needed to produce this lift.

In due time, a Collision Avoidance System (C.A.S.) was added to the two Maui radars. This was a computerized depiction of radar echoes, which displayed the course and speed of such targets on a separate large screen between the two originating radars. The C.A.S. similarly projected Potential Areas of Danger (PADs) wherein the computer had determined that a collision might occur should the Maui, and the detected ships (or other echoes), maintain their original courses and speeds. In short, the entire display was automatically plotted, updated and analyzed by the computer, relieving the deck officers of the plotting task. With the aid of this sophisticated addition, the behavior of squalls could be readily determined, such as any changes of their directions of motion and speed. Experience with the CAS left no doubt that rain squall echoes were reacting to some kind of emission or influence from the ship. The results naturally varied with latitude and Maui's course and speed, and the season of the year, which determined the direction of the etheric flows at that particular time: north to south in the northern hemisphere fall and winter, and south to north in the northern hemisphere spring and summer. Moon phases definitely ruled the intensity of the west-east etheric flow in temperate latitudes.

ETHERIC RAIN ENGINEERING ON THE HIGH SEAS

13 YEARS OF OBJECTIVE, CONCRETE, REAL-WORLD RESULTS

"THE MOSES EFFECT"

The SS Maui's Collision Avoidance display provides record of the typical "parting of heavenly waters" that occurs when the vessel approaches rain masses with its biogeometric devices functioning. "X" is the position of the Maui, at center screen. Screen shows ocean surface over 24 NM radius circle around the ship. Heading flasher (dotted line) shows ship steering SW for Honolulu from Los Angeles. Rain mass on both sides of ship has 'parted" for her passage, and open trench" through rain is clearly visible. Phenomenon has been observed literally hundreds of times during TJC's maritime experiments with control of local weather via etheric engineering.

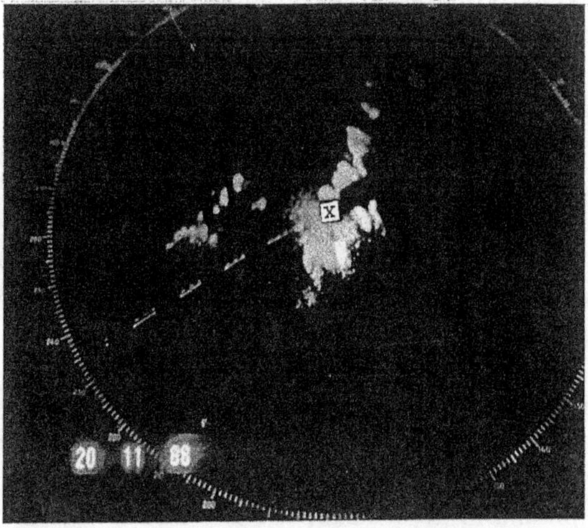

DAMMING UP ETHERIC FORCE FROM THE WEST

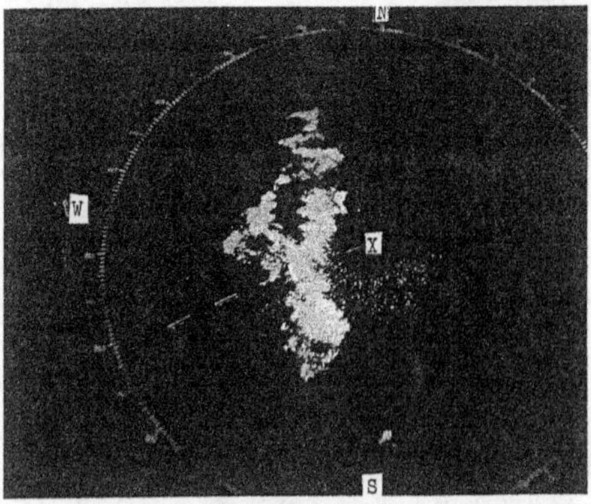

SS Maui (X) is one day outbound from San Francisco, California on her way to Honolulu. In the temperate zones of the earth, the prevailing ether flow is west to east. Maui's southwesterly course is shown by the dashes and spaces of her heading flasher, each dash and each space denoting 2 nautical miles. The barometer is 1022 millibars.

The biogeometric equipment aboard the SS Maui has been adjusted to oppose locally the regional west-to-east flow of etheric force, utilizing not only the translators, but also the 22-knot speed of the ship. Skill, experience and patience are needed to produce etheric drainage flow away from the ship, and directly into the advancing etheric flow from the general direction of magnetic west.

Where the ether flow is opposed in this fashion, the etheric potential rises, strongly attracting atmospheric moisture. The anomalous, engineered area of elevated etheric potential then becomes the target of lower potential flow approaching from farther west. Following the low-to-high flow law of etheric currents, this powerful flow soon raises the etheric potential of the accreting rain area to "lumination" or discharge point. Heavy rain then ensues, as in this radar depiction. This represents the etheric physics of the situation above.

Radar shows the SS Maui almost dead center of a monstrous rain buildup, well over 20 miles long and up to 8 miles thick. Sea clutter south of the ship verifies the southerly winds in this format, which have virtually no influence on the rain accretion — the latter being etherically controlled. Such control remains in place until the rain barrier surges past the damming influence exerted by the shipborne equipment. The rain formation then melts away, as the west-to-east etheric currents normalize after the removal of the local blockage caused by the ship.

Characteristically, in such situations as the above, the rain mass parts locally where the SS Maui approaches, so that the vessel commonly sailed through such rain masses as the above, in virtually rainless conditions with heavy precipitation on both sides. Common occurrence of this "parting of the waters" phenomenon, caused it to be nicknamed "The Moses Effect."

"Some day we will have full technical control of the ether," says TJC.

ETHERIC RAIN ENGINEERING ON THE HIGH SEAS

RADAR — NO METAPHYSICS, NO MAGIC, NO MYSTICISM

Colossal wall of rain, over 40 nautical miles long and up to 10 miles thick, advances on the SS Maui (X) from magnetic east — after being engineered into existence by biogeometric devices aboard the ship. The diagonal line of dashes is the ship's heading flasher, each dash and each space denoting 2 miles of over-the-ocean distance. Scope diameter is 48 NM. The ship's speed is 18 knots.

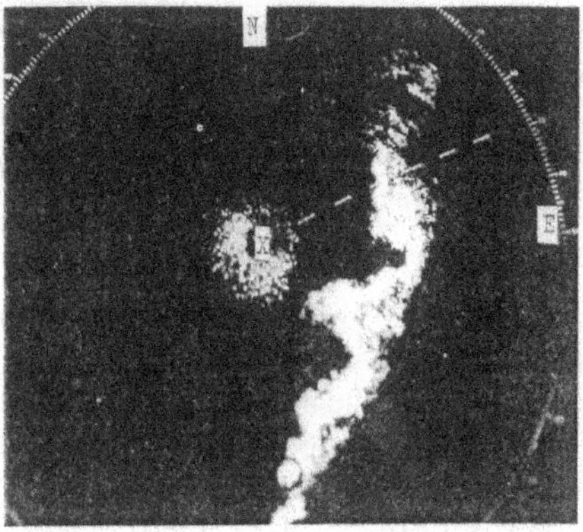

As the ship steers 070 degrees for California, the east-to-west flow of etheric force near Oahu, Hawaii, is being dammed up by engineering procedures, as the vessel moves. Damming this vast flow of force, causes a local rise in etheric potential. This in turn, causes lower potential etheric force from farther east to pile into the dammed area, obeying the "reverse" flow laws of the etheric continuum, and elevating local etheric potential. The elevated etheric potential powerfully attracts atmospheric water vapor, regardless of the barometer, which in this case was 1020 millibars.

Result is the creation, "out of nothing," of a vast rain line like this, containing millions of tons of water. Massive, continuous pushing by the east-to-west tropical ether flow, against the pulsatory etheric emissions from the shipboard apparatus, produces still more etheric excitation, with the whole system eventually driven to "lumination" or discharge potential, releasing torrential rain. Rising winds west of this engineered front show in the wide band of "clutter," or spurious sea return, around and mainly behind the SS Maui.

Photo #2 shows what happens once the engineered rain line has overwhelmed the Maui, removing the source of the damming effect that created this rain system. A scant 12 miles beyond the ship, the reflectivity and size of the rain mass have sharply diminished, and it has already shrunk from more than 40 to less than 27 miles in length. The whole rain mass has also slewed northward into a new orientation to the Maui, almost parallel to her course. At the Maui, a wind shift of almost 90 degrees has occurred, as denoted by the change in the sea clutter distribution around the ship. The slewed rain mass is continuing to draw etheric force from the Maui's etheric translators, but once west of the ship, no damming influence can be exerted in these latitudes, like that shown in the first photograph. The rain mass is also now geometrically out of tune with the shipboard translators, and will soon discharge itself out of existence.

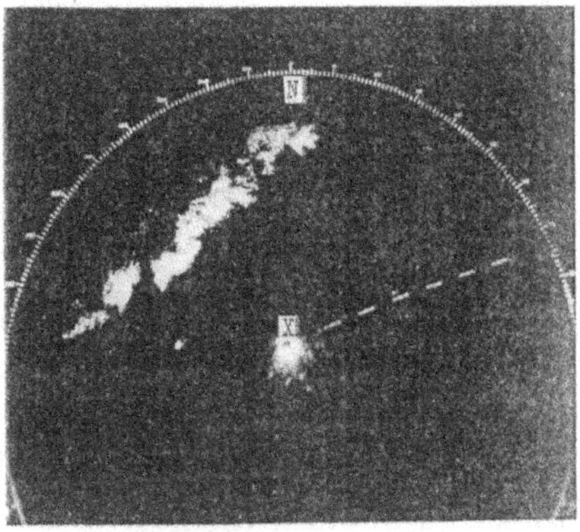

Scenarios like this were enacted literally hundreds of times during TJC's more than 300 ocean crossings aboard the SS Maui. Recorded often in daylight on time lapse video tape, the happenings were totally objective. In addition, radar provided continuous, irrefutable evidence of the validity of etheric flow laws, and of the physical presence of enormous tides of etheric force that are involved in the whole life of the earth. The larger lesson, according to TJC, is that the ether is technically accessible, and manipulable via the correct and appropriate geometry.

The water-powered rain engineering situation on the flying bridge eventually became a definite threat to the continuation of my privilege of having an installation there. The scuppers (drains) on the flying bridge were of inadequate design. Water from my equipment, or rain, or both, would therefore slosh over the retaining edge of the flying bridge as the vessel rolled slightly, and cascade down on to the bridge deck below and on down to other decks further below. This led to one of Commodore Orcutt's genius recommendations. He suggested to me that I did not really need the water to power my equipment. I ought to find a way to get rid of this water dependence. That would be technical progress because it would allow me to take the equipment anywhere and operate it without water. There had to be a way, he opined, and he expected that the Chief Engineer and myself, working as a team, could find that way.

This development was being quietly forced on me, just as had the downsizing of the massive tube batteries used ashore. The Chief Engineer of the Maui, Mr. Louis Matta, was among the most highly valued of my associates. He is a gentleman of noble motives, ethics and character. He had been interested for many years avocationally in sacred geometry, and was a student of Tai Chi. He took the ether in his stride, having long ago come across it as *chi*, in Oriental philosophy. Given Commodore Orcutt's gentle ultimatum, we got cracking on discussions, theories and endless sketching. Finally, we found a way that we theorized should work, and we fabricated what was needed in the ship's workshop. Lou Matta was, and is, an outstanding sheet metal artisan, as well as a top-flight marine engineer. A simple geometric assembly, not using water power, was put into action. Such a device required a certain familiarization period, but in a short time revealed that Commodore Orcutt had been right. We did not need the water.

Once again, radar proved out this advance objectively. Large rain buildups were possible without water grounding. The same general type of geometric unit proved useful on automobile roofs ashore, under suitable circumstances. Operating water-powered guns in cars had proved messy and functionally unsatisfactory. The airborne translators of today are descended in a line of provings from the first "waterless" rain guns fabricated by Lou Matta.

Years of tests and trials ensued. In their course, many variants of geometric rain engineering devices were tested and tried under the steadfast eye of the Maui's radar. One of Lou Matta's inspirations was to fabricate a translator that would theoretically "stir" the ether into vortices as we raced across the ocean. He designed a resonant chamber with two 13-inch projectors at 90 degrees to each other, so that they pointed skyway at 45 degrees to the vertical. The

45-degree angle was "locked" into the structure, which could withstand the gale force winds common on Maui's flying bridge. This simple assembly was then rotated by a small electric gear motor as the vessel moved. Given the appellation "Box Apache," this unit laid down parallel rain lines on each side of the Maui's course, which at times extended back behind the ship as far as 30 miles. Photographs of the Box Apache's rain pattern were taken directly off the Maui's CAS, and survive to this day.

Golden Section cones, usually four of them, were mounted with their apices pointing skyward, at an angle of 72 degrees from the horizontal. This type of unit was named a "Spider," because the mounting crossbars, upside down on the deck and minus the cones, resembled a huge aluminum spider. Rotating the Spider on the Maui, under way, surrounded the moving vessel with rain or near-rain conditions, which could be readily detected on radar. This circle of often massive rain echoes would move across the surface of the ocean synchronously with the Maui, an astonishing distribution of engineered rain, once again verified by radar's dependable eye. The Spider was obviously generating implosive etheric vortices steadily, to produce these anomalous rain patterns by "screwing" the rain into physical existence. In daylight, the existence of the vortex pattern was evident to the eye in the surrounding scenario, as well as on the radar screen. Directly above the ship, and moving along with it, was a shimmering blue ring of sky, carved right through the overcast. All around the horizon meanwhile, the rain clouds were being crushed down in a giant doughnut formation, 360 degrees around, 6-8 miles out from the moving ship. The entire vortex pattern moved synchronously with the ship, centered on the ship. Radar confirmed the objectivity of this scenario, examples of which appear in the publicly released video *Etheric Rain Engineering on the High Seas*. There are also, in this same video, examples of virtually all the functions mentioned in this summary of radar's vital role in opening us to understanding this engineering art. Especially noteworthy is the video sequence recording rain pillars descending into the ocean, with relative humidity at 45 percent, and a 10F differential between wet and dry bulb thermometers.

Perhaps the most significant part played by radar, overall, is the development of operating acumen, stimulated by radar's constant guidance. Some quarters might call it intuition or instinct. When the legacy of all these thousands of hours was converted to use on aircraft, that operating acumen found its most useful application: operating without radar if need be. Velocity in airborne etheric rain engineering had greatly increased over shipboard operations. The speed of all happenings was now raised. Effects were magnified by virtue not only of greater speed, but also by having the translator completely clear of the earth. Especially in tropical locales, reaction of the ether was rapid and could be violent. The operating acumen built up on the Maui and guided by radar's yield proved invaluable, especially as a safety factor.

With certainty it may be said that without the 14 years aboard the Maui, more than 300 crossings of the north Pacific, and those radars – in particular the radars – I would still be "down in the dark" where etheric rain engineering is concerned. The basis was laid by this work for the future, further investigations of the ether that will be carried out by formally qualified and trained scientists – using radar. These scientists will use radars of such sophistication, sensitivity and specialized capability, that the simple SSV (Ship to Surface Vessel) radars of the Maui would appear primitive by comparison. These capable men and women will, one day soon, lose their structural apprehensions of the ether, and overwhelm such fears with healthy curiosity about the etheric presence all around them. Funding will perhaps appear from those few in authority who are capable of conviction that a different and better world will be born of a comprehensive understanding and technological use of the ethers. Fortunately, on the Maui, we did not have to depend on retarded people, or waste our time trying to convince compulsive skeptics. Otherwise, radar would not have been able to deliver the enlightenment that was its gift to us, and to the whole world.

Go to RADAR GALLERY at http://www.rainengineering.com/
For proof, photos, good quality it seems, of WX engineering

Chapter 15

LIGHTNING IN ETHERIC RAIN ENGINEERING

An Overview, by Etheric Rain Engineering, Inc.

The engineering, stimulation, or enhancement of lightning is *not* among the purposes for which etheric rain engineering has been developed over the past three decades. At ERE of Singapore, we certainly wish our airborne operations could be kept free of lightning, because of many attendant dangers. Ignition of forest fires and disruption of commercial electricity supplies are especially undesirable. Lightning has nevertheless been concomitant to some degree with most etheric rain engineering operations since the 1970s. Our experience is that the ether is an integral component of lightning, and probably its fundamental source. This has been strongly indicated in more recent years, with the development of ungrounded translators (P-guns), and their use on aircraft in tropical and subtropical locales. ERE aircraft operating in Malaysia for example, have taken off in stable, fair weather with no changes forecast or indicated in any way, only to have thunder and lightning manifest profusely, right after becoming airborne with a P-gun aboard. This has frequently forced an immediate return to base – a chastening experience.

Considering that reactions like these are produced without any use of chemicals, electromagnetic radiation or electric power, but purely through simple geometry and lawful motion, something is crucially lacking in conventional lightning theories. That same something is in the forefront of airborne etheric rain engineering operations: accessing and manipulating the ether technically. Practical etheric engineering experience has taught that the ether is an intimate contributor to lightning, and perhaps the cardinal contributor. The ether is missing from conventional attempts better to understand lightning. Great diligence and devoted professional effort over many decades have attended conventional lightning research. ERE has the greatest respect for these largely unsung projects. Nevertheless, the physicist who chooses to dwell in the etherless universe wrongly derived from Einstein's mathematics, excluding the ether from such primal functions as lightning, is destined to have no definitive grasp of lightning phenomena. Excluding the ether only frustrates and muddles such understanding. Greatly enhanced comprehension of lightning phenomena will follow admission of the ether to the whole process.

Included in this gallery are lightning photographs where the deliberate intent of the engineer was actually to show photographically the connection between etheric rain engineering apparatus and lightning displays this equipment evoked. This original effort to demonstrate primitive "control" of lightning by having strikes appear in the center of the camera field, was made in 1978 – a risky venture with a twelve-foot long metal tube grounded into water, and sited atop an oceanfront building. The much-feared threat of being "zapped," fortunately never materialized, and confidence of personal safety grew. Practical experience led to the conviction that the immediate vicinity of the cloudbuster was a fairly safe place. Etheric energy seemed to drain away massively from the near proximity of the cloudbuster and not just via the cloudbuster tube. This drainage ran toward the site of the lightning strike, which was always remote from the cloudbuster. That site could be controlled to a limited degree by changes in the orientation of the cloudbuster.

***First Effort**. On 8 August 1978, an effort was made to forge an evidential photographic link between the Cloudbuster and lightning strikes. The operational site was atop an apartment building on the western San Pedro Bay shore, adjacent to the Los Angeles harbor breakwater. The Cloudbuster was Magnum 144, a 12 foot-long spiral wound metal tube, 12 inches in diameter. This unit, which looms in the extreme right foreground, is grounded via the building standpipe into the water organism of the earth. Magnum 144 was fed with atomized water. The lightning strike near the center of the photo hit in the ocean not far from Newport Beach. This evening's experience showed that location of lightning strikes could be controlled to a limited degree by the elevation and orientation of Magnum 144, and also intensified. This indicated that the ether had a fundamental involvement with lightning, confirmed by later experience.*

Build-up of etheric potential was fed by this "drainage." If the resultant elevation in etheric potential became sufficient, reaching what is known as lumination potential, lightning discharged the etheric potential in a process bio-energetically tantamount to the sexual orgasm. Lightning appeared to function primarily according to the basic "4-beat" orgasmic cycle discovered and described in the scientific works of Dr. Wilhelm Reich. The 4-beat cycle lies at the roots of life. This primary activity is accompanied by electrical effects as conventionally understood and comprehensively objectified. Vivid and observationally transfixing, these electrical effects mask the preceding primary activity.

Kindred experience on the high seas, by engineering vast rain lines containing thousands of tons of water via similar etheric "drainage" phenomena, convinced us that we were dealing with phenomena rooted in living processes. The legacy of Dr. Wilhelm Reich was of immense service in advancing our understanding.

The Photographs

The 1983 photographs were made using simple, short (34-36") large-bore tubes in water-grounded etheric projectors, not really worthy of being called cloudbusters. These devices were crude and inelegant, awkward to handle and control, and incredibly messy to use. The units were fed by blasts of atomized water. Operators were completely drenched in operating these units. Their virtue was the large output of etheric force that they made available, and their ability to project this force locally with serviceable directional control. Most of the photographs in this series, made at Los Angeles Harbor in California, reveal the upper tube of the projector in the lower center of the photo frame. Connection to the massive lightning strikes is therefore unequivocal, since many photographs were made utilizing the same method and geometry.

Hi Neighbor! *A 4am wake-up lightning strike slams into the ocean off Point Fermin, CA on 14 Aug. 1983. The upper barrel of the cloudbuster points south over the home of TJC's neighbor – a prominent L.A. criminal attorney – who was rousted from bed by the immense thunder from this and subsequent strikes. Note how the cloudbuster orientation "tracks" the lightning strikes in this series.*

Another Lucky Shot. *Right after "Hi Neighbor!" in same basic format, except that the cloudbuster has been moved eastward from original siting in order to continue production of lightning strikes. The rationale for the series of photographs was to show direct linkage of the cloudbuster to the lightning. "Power" source of the cloudbuster was a blast of atomized water.*

Two Strikes

Three Strikes

Four Strikes

In end photo, four massive lightning bolts strike simultaneously into ocean waters adjacent to the Los Angeles harbor breakwater, 14 August 1983, at 4am. The palm trees of Cabrillo Beach occupy the foreground. At the extreme right, the upper barrel of the cloudbuster used on this occasion to maximize the lightning, looms into the picture, but now heading due east. A poster-sized print of this picture hung in the officers' mess of SS "Maui" for many years Kodacolor 200 was the film used for this series.

The lightning photographs made during Operation Pincer II (July 1986) in southern California, permanently document the culmination of that complex project. Filed with engineering drawings prior to commencement with the National Oceanographic and Atmospheric Administration (NOAA), Pincer II was an attack on a century-old statistical barrier: the statistical rainlessness of Los Angeles in July. The result of Pincer II was the wettest Los Angeles July in 100 years. In order to achieve this etheric engineering triumph, the July monsoon that normally affects Arizona and especially the Phoenix area, had to be diverted over 200 miles from its normal path. A full account of the Pincer II project can be found elsewhere in this book.

Pincer II marked the advent in a pre-notified project of ungrounded, waterless, geometric etheric projectors. These "Flying H'" units, in addition to engineering July rains in L.A. not seen in a century, evoked southern California lightning on a seldom-seen scale. There were approximately 200 lightning strikes in the vicinity of Point Fermin, 22/23 July 1986, where the coastal Flying H was sited. This amount of lightning activity was probably unprecedented for this region. Lightning is relatively rare in Los Angeles.

THE APPROACHING STORM
Spectacular, cosmically-illuminated panorama in San Pedro Bay, as rain and lightning enter the area from the southeast. This was the coastal "pincer" of rain activity laid out in the engineering drawing filed with the American government (NOAA) before Pincer II commenced. Time is approximately 1 am, with no rain in the official forecast for 23 July.

Flying H in Action #1

Flying H in Action #2

Flying H in Action #3

Flying H In Action, from 23 July 1986, approx. 1:15am. In photo #1 the Flying H weather engineering unit rotates on an easterly bearing, with its main axle lying north and south. Four simultaneous lightning bolts strike into San Pedro Bay, providing a cosmic flashgun for this photo. The L.A Harbor Light is near the center, with Cabrillo Beach palm trees in foreground.

In #2, the rotating Flying H unit looms in left foreground, provoking a massive single bolt down from perhaps 30,000 feet. This bolt strikes the ocean just beyond the L.A. Harbor Light, which appears near the center of horizon line. The engineering purpose of this operating format was to divert rain accretions approaching from the southeast onto a course east of Point Fermin, where this Flying H is emplaced. The objective was to have the rain fall into the Los Angeles Civic Center official rain gauge, which it did, for the wettest Los Angeles July in 100 years.

In #3, another direct connection between etheric rain engineering and lightning is demonstrated during Operation Pincer II. Two simultaneous lightning bolts bracket the anchored cargo ship in L.A. Harbor. A compelling objective record of the pre-filed Operation Pincer II was made on National Weather Service radar. These government radar facsimile depictions irrefutably show rain formations moving in two "pincers" into Los Angeles from Mexico – 200 miles off their normal track into western Arizona. The radar record outpictures the engineering drawing filed with the U.S. government weeks before the operation commenced, and is part of the documentation of Pincer II (see next chapter).

The photographs made on this spectacular occasion are of two types. First, salon-quality photographs of lightning strikes, all of them right in the center of the frame, a verification that the engineer knew approximately where the strikes were to hit (see below). The second group of photographs are framed and sited directly to connect lightning strikes around Los Angeles Harbor to the Flying H units that evoked these displays (above). Nothing remotely resembling the lightning of Operation Pincer II has occurred since, a period of 16 years.

***"SALON" PHOTOS,** were taken from the engineer's roof, fifteen feet higher than the Flying H series of photos, this same night, 23 July 1986. There are therefore no obstructing wires to mar the beauty of Mother Nature in action. In every case, the "Salon" photos have the lightning strike in the center of the frame, or close to the center. The Flying H made it possible to hold lightning strikes to the same general area for several minutes, hence the near perfection of these images. Photos in this Salon series were eagerly sought by Los Angeles Harbor residents, and hang to this day in the offices of doctors, dentists and businessmen, as well as in numerous homes in the region. Capturing with a photograph the tremendous forces inherent in lightning is extremely difficult, because of the extreme brevity of lightning strikes.*

In the course of 34 years in the pursuit of etheric rain engineering, lightning has attended numerous operations and equipment tests on the high seas, as well as in Hawaii, Singapore, Malaysia, Costa Rica and on the deserts of California, Arizona and Utah. Lightning has synchronously accompanied the passage across American deserts, a mile or two abeam of moving "gun cars" equipped with ungrounded, geometric translators. Such happenings left no doubt of the connection of the lightning to the moving vehicle. Similar, cyclic lightning discharges were engineered to accompany an ocean-going ship during a hundred nautical miles of steaming. The strikes always appeared on the same bearing relative to the ship, at the same approximate distance, and at regular intervals. The "message" was always the same: lightning originates in etheric activity.

ERE is driven and guided by objective results. Our understanding of etheric functions and influences is in its infancy. Slight though our advances are in this revolutionary new art of etheric engineering, we are enjoined, by the dynamic nature of what we have discovered, to avoid freezing any of it yet into either doctrine or dogma. Trying to fit etheric functions into the mechanistic knowledge rubric that is based on denial of the ether's existence, is irrational. Since the physical existence of the ether is intimately connected with all life, the ether as a reality, and as a concept, must be allowed to flow freely into human consciousness for a long time to come. In this way, the ether itself will "tell" us of its wondrous properties and powers. This experience will undergird a functional new methodology for dealing technically with the ether, and especially for the ether's technological use. Using existing, accepted mechanistic laws to deny, "test" or deprecate the ether for the sake of protecting the mechanistic weltanschauung is a neurotic mistake. Blocking or banishing the ether from scientific consciousness, is the grossest of counter-evolutionary errors.

New theories and concepts appropriate to etheric phenomena will undergird the age that lies ahead. Compulsive skeptics will do well to remember that many "laws" now hallowed, will join the Phlogiston Theory in history within a few decades. At ERE, the intention is to successfully commercialize etheric technology for the first time, since the world inevitably accepts what it has to pay for. This is the ultimate seal of approval. Wider acceptance and technological use of the etheric continuum can follow from the break-in with etheric rain engineering. A new epoch looms before industry and commerce To help establish this new, life-positive era, based on the physical reality and technological use of the ether, the discipline imposed by *results* appears to be a safe beginning pathway. A century of scientific disputes and endless argumentation about the ether has led nowhere. There is much more behind lightning than familiar electrical forces.

Flying H in action during Operation PINCER II, 23 July 1986. Note how a smaller strand of lightning on the left side comes far into the foreground, connecting directly to the pair of horizontal electrical wires or the actual Flying H device itself – or both. This image also used on front cover.

Chapter 16

OPERATION PINCER II – 1986
Overview

July Rain for Los Angeles, California

In presenting an account of a successful etheric rain engineering operation of great complexity, information previously undisclosed concerning the preludes to Pincer II is essential. Especially is this true for thousands of persons worldwide, who will be making their first acquaintance with etheric rain engineering via this book. This applies both to intelligent laypersons and professional people. The technology employed in Pincer II is no longer used as a separate entity by ERE of Singapore, but is incorporated in current technology, including AEREO (Airborne Etheric Rain Engineering Operations). The principles employed in Pincer II are basic to the art, and likely to be enduring.

Background

The road to Pincer II was long and hard. Pincer II did not just happen, like a rabbit pulled out of a magician's hat, or lucky coincidence. Many shortfalls, pitfalls, embarrassments and failures accompanied progress along a lengthy empirical pathway. On that pathway there was no conventionally accepted, classical undergirding for the project. There were no authorities that we could consult for guidance. This was truly pioneering, self-financed and independent.

In scale and scope, Pincer II was a geographically large and ambitious venture – unequalled to this day. Pincer II climaxed a series of July rain engineering operations beginning in 1976, all with the same purpose. The objective was to conquer a well-established barrier: the statistical rainlessness of Los Angeles in the month of July. Successfully overcoming this century-old barrier, via etheric rain engineering, would provide fundamental evidence of the validity and value of etheric engineering.

Documentation

U.S. Federal laws provide that such weather modification operations be reported *in advance* to the National Oceanographic and Atmospheric Administration of the U.S. Government. By appending an engineering drawing to our Federal filing for Pincer II, we would show clearly how it was intended to bring rain into statistically rainless Los Angeles in July. U.S. Government radar fax maps would provide third-party confirmation and irrefutable objective evidence of where rain formations developed, and how they moved and thus brought rain into Los Angeles – were the project successful.

Simply expressed, we would define in advance what we intended to do, and how we intended to do it, using the official government filing procedure. We would then execute the project. Government measurements and radar maps would objectively confirm – or refute – our success.

Early Operations

In the period between 1968 and 1976, many exploratory operations via etheric rain engineering techniques were carried out, using for the most part, extremely simple "rack" units grounded into flowing water. These crude assemblies of 100 or more hollow tubes were not to be compared to the elegant cloudbuster designs of Dr. Wilhelm Reich, who devised the basic techniques involved. Our large and rough rack units nevertheless mobilized vast amounts of ether. Such units provided basic practical experience in manipulating the ether, and confirmed not only its objective, physical existence, but also its technical accessibility. These huge, crude installations were sufficiently powerful that their results "forgave" and compensated for our amateur ineptitude and inexperience. Evidence that the weather could be influenced over enormous distances and areas grew as the years passed. Confidence that such operations could be successfully mounted, developed naturally from consistent, practical participation in a new, scientifically-based art. This was on-the-job education. Most scientists denied the existence of the ether, and there had been squabbling over the ether by great minds for more than a century. The practical world was our venue and it was ruled by results.

Southern California was the scene of most operations. We became familiar at first hand with the regional weather and climate, in the particular way they are influenced by etheric flows and factors. This enhanced and illuminated existing, formal knowledge of the region. As our capabilities, equipment and operating acumen improved, we became able to engineer relatively astonishing conditions. The etheric continuum we were accessing was clearly of high power. The resultant effects were frequently so vast that they strained the credulity even of enlightened, well- disposed people who were convinced of the ether's reality. One came to think in states and even continents, while aware that in decades ahead, one would have to think on a planetary scale with etheric engineering. Hundreds of miles in distance, and thousands of square miles in area were involved in some of these primitive, pioneer operations.

IDENTICAL PRIMITIVE APPARATUS, IN THE 21ST CENTURY, CAN STILL READILY REVERSE DROUGHT IN MANY CLIMATES, WHEN DIRECTED BY AN EXPERIENCED RAIN ENGINEER. POWERFUL POLITICAL FORCES PLAN TO "PRIVATIZE" WORLD WATER RESOURCES FOR THE EXPLOITATION OF SUFFERING MANKIND.

THE EFFECTIVE RESPONSE TO WORLD DROUGHT IS TO TECHNICALLY ACCESS BILLIONS OF TONS OF PURE WATER THAT FLOAT ABOVE THE EARTH, USING ETHERIC ENGINEERING. PINCER II EXEMPLIFIES THIS APPROACH.

Moisture From Mexico

Establishing a large desert installation in Thousand Palms Oasis, about 15 miles northeast of Palm Springs, California, we grounded our apparatus into a 300 gallon-per-minute irrigation flow. With this incredibly simple rig, we found it possible in the searing summer months, by using etheric engineering techniques, to draw a large flow of humid air – and sometimes rain – from the head of the Gulf of California in Mexico, into the Coachella Valley region of the lower Mojave Desert. If the tubes at Thousand Palms were grounded into the irrigation water flow, and left untended, humidity in the lower Mojave Desert rapidly built up to near- intolerability for residents. Date crops verged on ruin in such anomalous, highly humid weather.

Brief ON/OFF applications of this simple technique left no doubt that it was the rack units that were producing the humidification. Through such experiences, the basic technique of damming up a fundamental etheric flow, and thus diverting it around a fixed base, was developed. Modifications of regional weather could be initiated in this way. Once proved out, the desert humidification experiments were not repeated, for obvious reasons. Among the operational possibilities arising from these tests was adaptation of the technique to an objective goal. Could we get the summer south-to-north flow of etheric force coming out of the Gulf of California in Mexico to curve northwest into Los Angeles, before turning away northeastward in its normal fashion? Humid air from the region around the head of the Gulf of California, impelled by a south-north ether flow, normally feeds a curving summer "monsoon" moisture flow north and east to the Phoenix area in Arizona. Our intention was to shift this moisture-laden etheric flow more than 200 miles westward from its normal track, by means of etheric engineering. The distances involved, and the distortions we were to engineer into established weather patterns, were daunting. Amateur enthusiasm has the power to minimize such formidable obstacles. We already had abundant evidence of what even the simplest water-powered etheric projectors could do, and no skeptical academician had the power to verbalize away this practical experience.

1976-1986 Run-Up

Every July from 1976 onward, we tackled some aspect of the Herculean task of engineering rain into Los Angeles. Practical experience proved over the years that with existing techniques, knowledge and equipment, we could come close to achieving our goal, but something was still missing. Definitely needed was an additional operating base well east of San Diego, California. Such a base was required reliably to draw the summer south-to-north ether flow out of Mexico and bend this moisture-bearing ether flow west of the coastal mountain range in southern California. Only then could the moisture-bearing ether move up the coast, west of the mountains on its northward passage to Los Angeles, guided in its final phase by our base on Point Fermin, the southernmost tip of Los Angeles.

From early Thousand Palms Oasis experiments previously mentioned, we knew that it was possible to engineer rain into the Coachella Valley, but the marked tendency of such summer moisture was to drift rapidly eastward into Arizona, pushed by the fundamental west-to-east flow of ether that prevails in the northern temperate zone. The most promising route for Los Angeles rain originating in Mexico was the coastal route, as previously described. Each July advanced our experience and capabilities further, but progress was slow. Around 1984, we secured the welcome cooperation of a forward-looking San Diego attorney and his wife, who was also an attorney. Surprisingly, both of these outstanding people had heard in law school of Dr. Wilhelm Reich and his legal battles. Both also were familiar with the career of the famous Californian rainmaker, Charles Mallory Hatfield. The latter's operations had flooded San Diego in 1914. A memorial to Hatfield stands today at the Morena Dam, not far from our benefactors' home.

Hatfield Flat

We christened the lawyers' mesa from which we could henceforth make our July efforts, "Hatfield Flat." From1200 feet above sea level and 16 miles east of San Diego, Hatfield Flat looked directly south into Mexico.

There was no doubt about the tremendous leverage possible from Hatfield Flat on Mexican moisture. Early operations quickly verified, via numerous all-time record July humidity levels in the San Diego region, that we were finally closing in on our longtime objective. Thunderstorms, lightning and rain sometimes battered Hatfield Flat, on one occasion for several hours. This proved the tactical value of the base. Storm damage to our warm-hearted benefactors' property was accepted with grace and good humor.

Flying "H" Units

The development at sea in 1984-86 of Flying H geometric units that did not require water grounding, further advanced our project. Great flexibility came with the Flying H (see also page 190). A water ground was not required, and they introduced rotary motion into the technology. At this time, Mr. Gino Segreti, a longtime resident of Desert Hot Springs, north of Palm Springs, kindly provided access to his property. Known as "Fort Zinderneuf," the Segreti property was ideally located for the purposes of Pincer II. Gino became a beloved, trusted associate and effective operator. He was for many months a shipmate aboard SS Maui. He developed his enthusiasm for rain engineering because of what he saw happen on the high seas, during shipboard experiments. These activities included the first marine use of the Flying H prototype.

The base in Desert Hot Springs made further use of Thousand Palms unnecessary. The new base, 10 miles north of Palm Springs, was much closer to the Banning Pass – which connects the lower Mojave Desert to the Los Angeles Basin. The new geometry of this situation opened the prospect of using the strategically placed Fort Zinderneuf base to push moisture entering the Coachella Valley from Mexico, through the Banning Pass and into the Los Angeles Basin. Experimentation with the Flying H units, and the new bases at Hatfield Flat and Fort Zinderneuf, led to high optimism that the goal of July rain in Los Angeles was within reach.

Splitting The Flow

Much practical experience now suggested that by splitting the summer flow of ether out of Mexico, an effective "pincer" effect could be created.

One pincer would be triggered from Hatfield Flat, following the line of the coast northward to Los Angeles. The other, interior pincer in central southern California would follow the natural "waveguide" provided by the sunken terrain in which lies the Salton Sea. This large inland body of salt water essentially attracted any ether flow moving northward out of Mexico, guiding the ether flow along the massive trench that holds the Salton Sea, all the way to Fort Zinderneuf, 10 miles north of Palm Springs. "The Fort" had virtual line-of-sight to the Salton Sea from 1200 feet above sea level. Fort Zinderneuf equipment could block further northward progress of any rainy formations and steer them into and through the Banning Pass to enter the Los Angeles Basin from east of that city. These were the two arms of the Pincer. A supplementary base was activated at 3400 feet above sea level on the north side of the Banning Pass, to facilitate and ensure the passage of any rain through the Pass.

This was the fundamental etheric engineering concept that led to the Pincer operations. There were two operations so named. The first Pincer, in 1985, fell short of success, but proved the basic conception to be correct and feasible. Changes and adjustments were made for Pincer II, scheduled for July of 1986, and the success of this operation is now history. The engineering information presented in this background summary should dispel any notion that Pincer II was an off-the-wall accident. Our thinking had been expanded along with our experience over the previous 16 years, including a new factor that had eluded us for many years.

Influence of the Moon

The extended period of experimentation in the southern California region eventually made us aware of an especially vital etheric component to success – the moon. Practical provings convinced us that the fundamental west-east flow of ether in the northern temperate zone was keyed to moon phases. The chief effect observed was that the west-east ether flow peaked at full moon, and then collapsed in the ensuing period, before recycling itself to maximum flow at the next full moon. This west-to-east etheric "push" was primarily responsible for the normal south- to-north summer monsoon flow of moisture-bearing ether from Mexico, curving northeastward to the Phoenix area, once this flow moved clear of the head of the Gulf of California.

The west-east ether flow, and the south-north ether flow of the northern hemisphere summer, are separate functions of the ether, but to some degree they mutually influence each other. In the case of the planning for Operation

Pincer II, full moon and the days immediately following would be the critical period when the possibility of "bending" the rain-bearing etheric flows out of Mexico, more than 200 miles to the northwest of their normal track, would be greatest. Immediately following the full moon was the time for us to make a maximum effort with Pincer II. There would be minimal eastward shoving of the northward-moving, rain-bearing etheric flows from Mexico. Most of the month, there was too much eastward shoving on these flows for them to reach Los Angeles. The statistical rainlessness of Los Angeles in July had been identified as a complex function of regional etheric flows, including cyclic lunar influence on those flows.

This lengthy empirical and theoretical background to Pincer II has never been previously disclosed. Considerable sacrifice in pioneering this work was made by the late Irwin Trent, who toiled and tithed for over 30 years in silent dedication as a dear friend and co-worker. The late Dr. James O. Woods gave of his considerable best in advancing this work, right up to the time that ill-health overtook him in the middle 1980s. His passing ended a memorable association of more than a quarter century. These two gallant gentlemen were indispensable contributors to Pincer II.

In 1987 I wrote the following account of Pincer II, which has been abridged here, in view of the extensive background detail in the foregoing. In the abridgement, the flavor of a recently completed operation has been retained.

OPERATION PINCER II
July Rain Engineering, Los Angeles, California 1986
by Trevor James Constable

The statistical chances of measurable rainfall at Los Angeles Civic Center in July of 1986 were "unequivocally zero" according to National Weather Service statisticians, quoted by the *Los Angeles Times* on 24 July 1986. Pincer II was a successful attack on this statistical barrier, using etheric energy technology and methods. Ten days prior to the project's commencement, I filed the legally required Initial Report with the National Oceanic and Atmospheric Administration (NOAA), the responsible Federal agency. A copy of the Report appears herewith (next page). An engineering drawing filed with that Initial Report, is reproduced in simplified form here as well.

TO: Atmospheric Programs Office, RD2
National Oceanic and Atmospheric Administration
Rockville, Maryland 20852

NOAA FORM 17-4 (4-81) — U. S. DEPARTMENT OF COMMERCE, NAT'L OCEANIC AND ATMOSPHERIC ADM.

INITIAL REPORT ON WEATHER MODIFICATION ACTIVITIES (P.L. 205, 92ND. CONGRESS)

1. PROJECT OR ACTIVITY DESIGNATION, IF ANY: Operation PINCER II

2. DATES OF PROJECT
   a. DATE FIRST ACTUAL WEATHER MODIFICATION ACTIVITY IS TO BE UNDERTAKEN: 1 July 86
   b. EXPECTED TERMINATION DATE OF WEATHER MODIFICATION ACTIVITIES: 31 July 86

3. PURPOSE OF PROJECT OR ACTIVITY: Research. Rain, L.A.Basin. Rain So. Calif. Nat. Forests (Anti-Fire) & Humidity Elevation, San Diego County.

4.(a) SPONSOR			4.(b) OPERATOR		
NAME: TJC-ATMOS INC.			NAME: SAME		
AFFILIATION: Independent Consultants		PHONE NUMBER: 213 833 4260	AFFILIATION:		PHONE NUMBER:
STREET ADDRESS: 3726 Bluff Place			STREET ADDRESS:		
CITY: San Pedro	STATE: Calif	ZIP CODE: 90731	CITY:	STATE:	ZIP CODE:

5. TARGET AND CONTROL AREAS (See Instructions)

TARGET AREA		CONTROL AREA	
LOCATION: see attached engineering drawing	SIZE OF AREA: SQ. MI.	LOCATION: This concept is not applicable to these methods.	SIZE OF AREA: SQ. MI.

6. DESCRIPTION OF WEATHER MODIFICATION APPARATUS, MODIFICATION AGENTS AND THEIR DISPERSAL RATES, THE TECHNIQUES EMPLOYED, ETC. (See Instructions) Further developments of juxtaposed geometric forms for direct insertion into and manipulation of, the primary energy (etheric) continuum that underlies the atmosphere. Apparatus employs no electric power other than that for physical rotation of "Flying H", fixed base units. The latter are sited at 4 points as per attached drawing. Additional mobile units are carried in two "gun cars" for operation while traversing highways. NO CHEMICALS, NO RADIATION employed anywhere in this project.

7. LOG BOOKS: Enter name, affiliation, address, and telephone number of responsible individual from whom log books or other records may be obtained.

NAME: TREVOR J. CONSTABLE		
AFFILIATION: PRESIDENT, TJC-ATMOS		PHONE: 213 833 4260
STREET ADDRESS: 3726 Bluff Place		
CITY: San Pedro	STATE: CA	ZIP CODE: 90731

THIS REPORT IS REQUIRED BY PUBLIC LAW 92-205; 85 STAT 735; 15 U.S.C. 330b. KNOWING AND WILLFUL VIOLATION OF ANY RULE ADOPTED UNDER THE AUTHORITY OF SECTION 2 OF PUBLIC LAW 92-205 SHALL SUBJECT THE PERSON VIOLATING SUCH RULE TO A FINE OF NOT MORE THAN $10,000, UPON CONVICTION THEREOF.

8. SAFETY AND ENVIRONMENT

☐ YES ☒ NO Has an Environmental Impact Statement, Federal or State been filed? ... applicable.

☒ YES ☐ NO Have provisions been made to acquire the latest fo... ...ational Weather Service, Forest Service, or other... ...If yes, please specify on a separate sheet.

☒ YES ☐ NO Have any safety procedures ... or suspension of operations, monitoring methods, etc.) and any envir... to the possible effects of the operations) been included in the operation... please furnish copies or a description of the specific procedures and guidelines.

OFFICAL ADVANCE NOTICE TO NOAA

9. OPTIONAL REMARKS (See Instructions. Use Separate Sheet.)

NAME: TREVOR J. CONSTABLE		CERTIFICATION: I certify that the above statements are true, complete and correct to the best of my knowledge and belief.	
AFFILIATION: TJC-ATMOS INC.		SIGNATURE: Trevor Constable	
STREET ADDRESS: 3726 Bluff Place		OFFICIAL TITLE: President	
CITY: San Pedro	STATE: CA	ZIP: 90731	DATE: 20 June 86	PHONE NUMBER: 213 833 4260

NOAA FORM 17-4

Operational Plan for PINCER II

This simplified version (next page) of the detailed engineering drawing submitted in advance to the U.S. government (NOAA), shows how it was proposed to engineer rain into Los Angeles in July 1986.

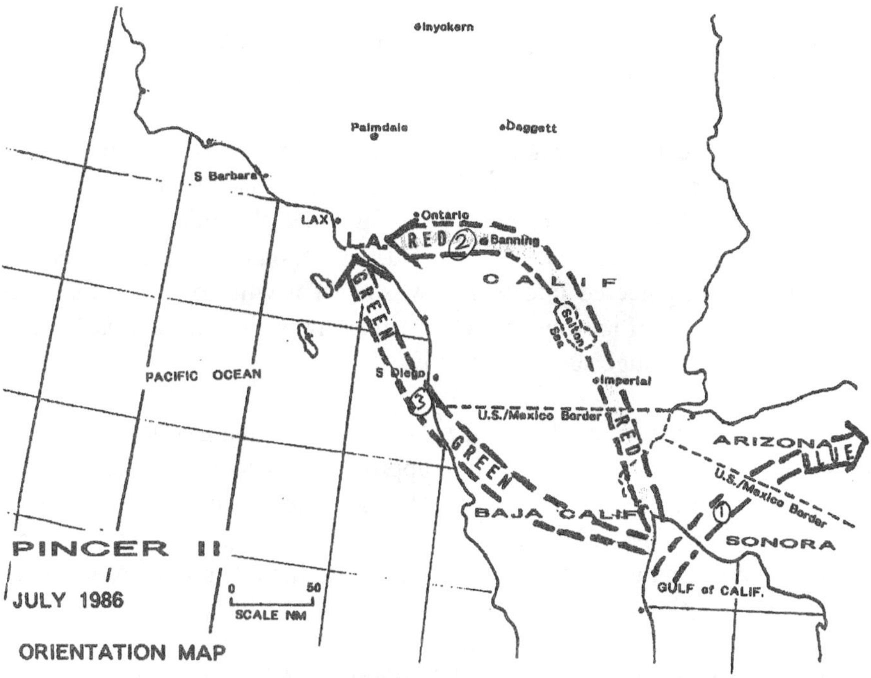

BLUE flow (pathway #1 in black & white chart) is the normal July path of moisture from the Gulf of California into Arizona, producing the Arizona "monsoon" season.

RED flow (pathway #2) is an engineered diversion of the blue flow. The engineering causes moisture to move, as shown on the map, into central Southern California. Bases at "Fort Z" and Banning, "bend" this flow through the Banning Pass into the L.A. Basin and to L.A. Civic Center. This is one arm of the "Pincer."

GREEN flow (path #3) is the second arm of the "Pincer," initiated by the base at Hatfield Flat, east of San Diego. This engineering induces moisture northwest from the Gulf of California and from Baja California, to flow up the coast toward Point Fermin, the southerly tip of Los Angeles. The Point Fermin base ensures that the rainfall moves northward to L.A. Civic Center.

The drawing shows the precise routes by which it was intended to engineer moisture from Mexico into the Los Angeles Basin. A highly spectacular and notably unforecast thunder and lightning show accompanied the unforecast rain that made July of 1986 the wettest Los Angeles July in 100 years and the second wettest July of all time. The moisture arrived via the pre-specified routes, as the accompanying National Weather Service radar fax maps objectively reveal. Pincer II broke the statistical barrier.

About 16 years of avocational work preceded this overnight success. Bringing rain to Los Angeles city in July requires that the normal, natural passage of Mexican moisture out of the Gulf of California northward and northeastward into Arizona, be diverted. An anomalous near-90 degree bend must be engineered into that flow, so that it will go some 250 miles northwestward out of its way. If such engineering is absent, the Los Angeles Civic Center rain gauge stays dry in July.

Our group today views the weather as a geometric living structure. Control is available to anyone wise enough to approach the task minus the biases of mechanistic meteorology. Patience is also a prime requirement in grasping and using what is essentially a bio-geometric technology. Planet Earth is a living organism, naturally reluctant to respond to the technology of chemical or electromagnetic insult.

Four operating bases were used in Pincer II. Sited at Hatfield Flat (16 miles east of San Diego); at Fort Zinderneuf in Desert Hot Springs; on the Banning Bench in the Banning Pass; and on Point Fermin (the southernmost tip of the City of Los Angeles); these bases were manned by trusted associates – Trent at Hatfield Flat, Segreti at Fort Zinderneuf, Norgaard in the Banning Pass, and myself at Point Fermin.

Pincer II derived its name from the engineering intention to bring moisture into the L.A. Basin via two routes designed, like a pincer, to converge on L.A. Civic Center. The main effort would be via Hatfield Flat and San Diego, up the coast on the ocean side of the coastal mountains, passing east of Point Fermin and dropping into the rain gauge at Civic Center. The secondary arm of the pincer was via the Salton Sea and Banning Pass into the L.A. Basin from its eastern end.

Successful engineering of this kind would unfailingly produce rains in the National Forest areas. Such rainfall was accordingly made a secondary goal of Pincer II, and was so stated in advance, in the Initial Report filed with NOAA on 20 June 1986.

Hatfield Flat, Fort Zinderneuf and Point Fermin were each equipped with the new Flying H units, developed in 1985-86 in dynamic tests on the high seas. The pivotal base at Point Fermin was also to test operationally the newest development of geometric weather guns: the Apache. This is a unit that works vertically, and induces implosive etheric reaction. The gross effect of such a device is the appearance of a small, local, low-pressure system, which usually migrates. In maritime mobile tests, it has been possible with sea-going Apache units to inject miniature lows into the massive eastern north Pacific high, and to have these anomalous formations actually appear on official weather maps and fax charts.

Fixed-base operations were supplemented during the project by the use of two "gun cars" on the southern California freeways. Irv Trent operated extensively between San Clemente and San Diego, on a stretch of highway lying mainly in the critical southeast-to-northwest direction. Several years of operating gun cars has conclusively proved that they can exert regional effects under conditions favorable for their employment. In more recent times, growing freeway congestion everywhere has reduced their effectiveness, velocity being a key parameter in their functioning. Efficacy of the Flying H units more than made up for reduced effectiveness of the gun cars.

Flying H units are able to "trigger" etheric flows, and to exert substantial diversionary influence upon them once started. They can also act as local dams to an etheric flow. This can result in a dramatic rise in etheric potential "behind" the dam, encouraging rain. Flying H units will boost lightning to maximum intensity under storm conditions. Successful use of this apparatus does not depend simply on its acquisition, but rather, on comprehensive knowledge of etheric forces and experience in their manipulation. My own empirical experience dates from 1957, and has involved thousands of hours of such involvement.

Pincer II went through three distinct phases, in the course of which substantial anomalies were engineered into southern California weather. Phase 1 was the 4th of July weekend. Forecast to be hot, a very cool weekend ensued, at the end of which rain had fallen in cities as widely separated as San Diego and Pasadena. Phase 2 ended on 15 July 1986 with a two and a half hour thunder and lightning storm in San Pedro Bay, east of Point Fermin. The display was spectacular (see Lightning chapter in this book), and was enhanced by the Flying H on Point Fermin. Colored photographs were made of these events, with the camera shooting right through the field of the rotating Flying H. Rain resulted in Long Beach, contra-forecast.

Phase 3 was the critical main effort, keyed to collapse of the west-east etheric flow after the full moon on 21 July 1986. Moisture was entrained from Mexico on the 22nd, via both arms of the Pincer. A mass of moisture billowed along the interior (RED #2) Pincer route, and thrust into the L.A. Basin via the Banning Pass, exactly as specified in the pre-filed engineering drawing. On the coast, initiation of Apache operations, in this period of west-east stagnancy, generated a small low off the coast of Los Angeles, which remained quasi-stationary. All units in concert brought moisture up on the GREEN #3 ROUTE east of Hatfield Flat, and then offshore west of the coastal mountains. The rapidity with which all this was consummated on the night of 22/23 July 1986, caught the National Weather Service and the TV forecasters flat-footed. Nothing of the spectacular events that transpired in the early morning hours of 23 July was forecast.

National Weather Service radar in Palmdale, California was sharply on its monitoring job. Passage of all moisture in the way described – past Point Fermin and on into L.A. Civic Center – is recorded forever in their faxed radar summaries. Some of those radar fax maps accompany this presentation. The radar maps showed our engineering drawing brought to life in the real world.

The lightning displays were the most spectacular seen in southern California in living memory. The greatest number of lightning strikes occurred around Point Fermin – approximately 200 in number. (Photos of this lightning activity with the Flying H can be seen elsewhere in this book). The booming light show stunned southern California, since no forecast or other intimation of such a violent event had been given on the weather segments of any 11 pm TV newscasts.

While events in the L.A. Basin seized local attention, including the wettest July day in 100 years, our attention was also on Segreti's operations from Fort Zinderneuf in Desert Hot Springs, supported by the Banning Pass installation. Only the wild pyrotechnics in L.A. could have outclassed what went on as a result of the RED #2 desert pincer. A thunderous light show took place against the face of Mt. San Jacinto, as scores of strikes slammed into its flinty slopes. More than an inch of rain fell in La Quinta. Hail appeared in the streets in the daytime, in July, in Palm Springs, Palm Desert and also in the Banning Pass. Hail there was marble-sized and noisy. Rain and hail drove Banning operator Norgaard indoors.

Pincer II produced sufficient rain in National Forest areas to overpower a number of stubborn forest fires. Some unofficial gauges reached four inches during the month of July, 1986. Combustibility of the national forests declined sharply. The secondary goal of the Pincer II was thus decisively achieved with significant public benefit.

The statistical facts of Pincer II are found in *Climatological Data, California*, July 1986, Volume 50, No. 7. In the two statistical divisions covering southern California, there are 100 official stations.

72 OF THE 100 STATIONS RECORDED RAIN IN JULY OF 1986.

Notable rains for the driest month of the dry season, were Anza 3.20", Cuyamaca 2.15", Idyllwild 1.73", Iron Mountain 1.47", Niland 1.50".

.18" of rain at L.A. Civic Center, the project bullseye, made July 1986 the wettest Los Angeles July in 100 years.

More than a century of official records provided invaluable statistical control. Pincer II showed for the first time in history that a complex etheric rain engineering operation could be designed, depicted in an engineering drawing, filed with the government in advance, and then executed successfully.

Control of basic natural forces has been a primal dream of the human race. Such control was clearly exerted in Pincer II, over a region of approximately 40,000 square miles. Humanity is being shown in this seminal work that already some "laws" of classical physics are apocryphal. The earth's tomorrows, if they are to be at all, will be characterized by constructive, harmonious and cooperative working with Nature's laws. This was demonstrated by ordinary men aware of such laws and minus large financial resources. Thus will we heal, sustain and transform the earth – Pincer II heralds the new ways.

Official National Weather Service facsimile maps of regional radar echoes, complete with irrefutable official time signatures, as follow, show that the "Pincer" was engineered exactly as advised in advance – weeks before the actual operation.

RED flow and RED Pincer came first, on the late afternoon and early evening of 22 July 86. The L.A. Basin is full of rain by 7.30 pm. The RED Pincer has closed on L.A. Civic Center.

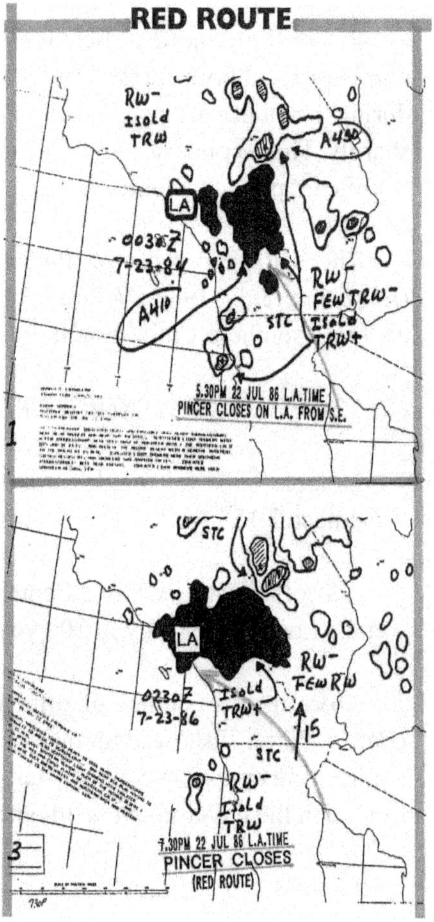

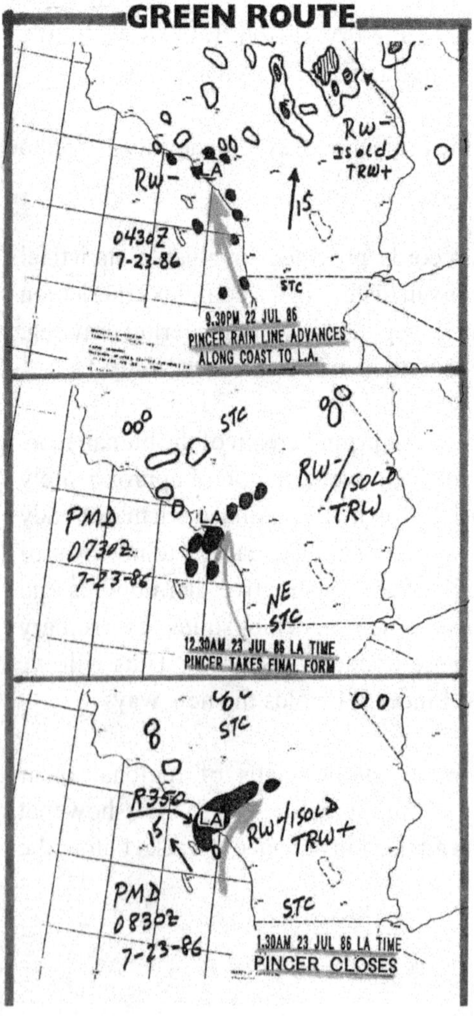

GREEN flow began showing on radar at 9.30pm on 22 July 86, spreading from south to north up the line of the coast. By 12.30am on 23 July 86, GREEN Pincer has reached Point Fermin and beyond. By 1.30am, the City of Los Angeles has been enveloped from the south and the GREEN Pincer has closed.

Chapter 17

THE ETHERIC FORMATIVE FORCES

Excerpt from *The Etheric Formative Forces in Cosmos, Earth & Man* by Guenther Wachsmuth, 1932

There are altogether seven etheric primal forces, formative forces, active in the cosmos; of these, however, only four reveal themselves in the space-and-time processes of our present phenomenal world. In what follows, therefore, we shall deal only with these four etheric formative forces.

Rudolf Steiner's Anthroposophical science designates these four kinds of ether as:

Rudolf Steiner

Warmth Ether
Light Ether
Chemical Ether (or Sound Ether)
Life Ether

The four kinds of ether may be classified in two groups, and this distinction is of fundamental importance for the understanding of all that is to follow: The first two, Warmth Ether and Light Ether, have the tendency to expand, the impulse to radiate out from a given central point; they act centrifugally; whereas the other two, Chemical Ether and Life Ether, have the tendency to draw in toward a centre, the impulse to concentrate all in a given central point; their action is suctional, centripetal. This polarity of the two ether groups – the centrifugal, radiating, self-expanding will, and the suctional, centripetal will to draw inward, to concentrate – is an ultimate elemental principle lying at the bottom of all natural phenomena.

Warmth Ether tends towards the spherical form. If it were merely a conveyer of "motion," then it could in turn call forth only motion in a substance-medium in which it works.

Since, however, the tendency to create spherical forms is inseparably linked with its action, therefore it calls forth, wherever it enters into Nature and is not obstructed in its action, spherical forms. We are here dealing – and this must again and again be emphasized – not with abstract dead oscillations of unknown origin, but with concrete formative forces.

The second ether state is that of Light Ether, or, more simply, of that which is given to the physical perception of man as "light." As Lenard says, light gave us the first intimation of the existence of ether, and he thinks "Light is undoubtedly a transverse wave motion: that is, in a beam of light and perpendicular to its direction – never merely backward and forward displacements in the same direction with the beam, as is the case in sound waves – there are present periodically shifting states. Optical researches by no means recent – for instance, those in regard to polarization of light, have already shown the transverse character of light waves. In the course of time we have learned to recognize still other ether waves which are invisible: ultra-violet, ultra-red, and electric waves; but these as a group have the same characteristics as light waves, differing only in their lengths." That the "characteristics" are similar, the lengths different, may satisfy us so long as we are testing in a one-sided and arbitrary fashion the quantitative-mechanical action in the substance medium; but in this way we learn nothing whatever in regard to the natures and the concrete distinctions of the different kinds of ether. The Light Ether to which we refer, which calls forth for the human eye in the manner to be explained later the phenomenon of light, does in fact induce among other things a transverse oscillation; but in addition to what has been said above we must add that this occurrence describes the figure of a triangle, so that Light Ether, as we shall see, when it can exert its effect unhindered in Nature, also produces there triangular forms, whereas Warmth Ether produces spherical forms. We may say, then, that an oscillation, a form which is caused by Light Ether in a substance-medium, takes the shape of a triangle.

The third ether is Chemical Ether, Tone or Sound Ether. Its forces, that is, cause the chemical processes, differentiations, dissolutions, and unions of substances; but also – though, as it were, through activities in another field – its forces transmit to us the tones perceptible to the senses. The inner kinship of these two spheres of action will be clear to us from the phenomenon of Chladni's sound-forms. For it is tone which causes the uniting together, the orders and forms, of substance and bodies of substance. "That which the physically audible tone produces then in the dust is happening everywhere in space. Space is interpenetrated by waves produced by the forces of Chemical Ether," which, in the manner of the Chladni dust figures, dissolve and unite substances. But Chemical Ether has in reality "a tone-and-sound nature of which sensible sound, or tone heard by the physical ear, is only an outward expression: that is, an expression which has passed through air as a medium."

We must establish the fact that tone and chemical processes are to be attributed to the same ether in the manner explained.

Chemical Ether, when it can exert itself unhindered in Nature, produces, as we shall be shown concretely, half-moon forms.

In contrast with the expansive kinds of ether – warmth and Light Ether – Chemical Ether, as we have said, tends in its action to be centripetal.

It may also be proved that the phenomenon of cold is one of those attributes which are to be ascribed to Chemical Ether, a fact which is essential for an understanding of the relation between processes of cold and of contraction.

The fourth ether is Life Ether. It is phylogenetically the most highly evolved ether, and therefore in its qualities most varied and complicated. It is that which is rayed out to us, among other things, from the sun and then modified in its action by the atmosphere of the earth. Life Ether, together with Chemical Ether, belongs to the group of suctional forces – those which tend to draw inwards. We shall also be able to prove its relation to that which is called "gravitation" and to the phenomenon of magnetism.

Its form-building tendency, when it can exert its effect unhindered in substance, leads to square shapes, expressed, for instance, in crystallizing salt.

Chapter 18

PROJECT TANGO

Rain Engineering in the Singapore Dry Season, July 1988

Primary energy rain engineering experiments conducted by me off Singapore Island in February of 1988 developed techniques for engineering rain, virtually at will, during most of the Singapore year. Later tests after my departure, carried out by my Singapore associates, fully confirmed the practicability of the February developments. Singapore's dry season, with sharply reduced rainfall in June/July/August, remained an unanswered technical question, with it's particular set of conditions. Project TANGO was accordingly organized by our company as a private research venture to develop primary energy rain engineering techniques for the dry season in Singapore.

Successful results would be commercially applicable in other equatorial and subtropical environments bedeviled by dry-season difficulties. This U.S.-developed technology has been exported during America's own "Big Dry," as a result of prohibitive liability lawsuit risks within the USA.

Concomitant politico-legal problems are less amenable to control than the weather, and make it impossible for the pioneers to place this work in the service of the American people.

The latter two weeks of July were selected for TANGO as roughly coinciding with the middle of the dry season. Certain personages of unquestioned integrity in Singapore were advised in advance of the project. All were externes to our company. The main weather engineering unit, approximately the size of a human head, was placed aboard a 46-foot Hatteras cabin cruiser. Ancillary, fixed-base equipment was installed atop a 31-story building on the southerly shore of Singapore Island. An Apache vertical unit was operated at corporate HQ in the Loyang area. West of Changi airport.

Marine mobile operations commenced 19 July, 1988. By 22 July, 1988, an effective operating method had been worked out. The afternoon radar plots from Changi weather radar show the effects on the region of the primary energy engineering stratagems employed. The plots, between 3 pm and 6 pm, show the diminution of the accretions as the boat headed back from Indonesian waters toward Changi, followed by the rapid rise of rain formations as our "T" technique was employed due east of Singapore Island, between 5 pm and 6 pm.

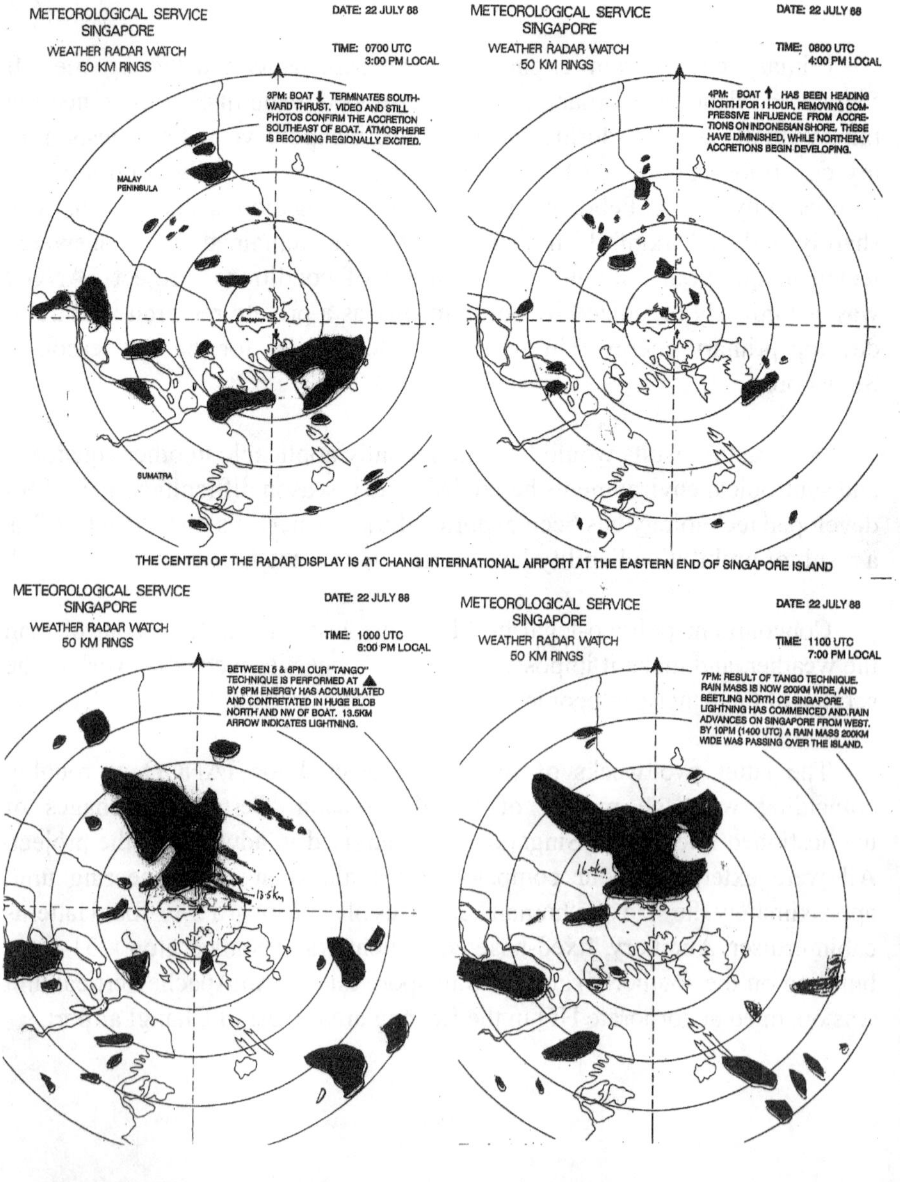

The primary energy methods employed are highly proprietary, but earlier disclosures on commercially released videotapes contain numerous examples of engineered rain formations becoming *self-reinforcing*. This active state was now engineered into the dry season conditions of Singapore and southern Maylaysia. A protracted thunder and lightning barrage ensued that evening, which included numerous strikes within 200 meters of the Apache unit at Loyang. A stout tree at the entrance to corporate HQ succumbed to a combination of lightning and gusty local winds.

One of the distinguished Singaporians who had been apprised in advance of our operations, received forceful proof of their efficacy. Storm winds ripped an expensive canopy from one of his business enterprises. Singapore is no stranger to violent tropical weather, but unforecast thunder and lightning in the middle of the dry season is unusual.

The ensuing disturbances on 23 July 1988 were kept active by judicious use of the fixed-base units. The situation was then permitted to subside, allowing for a national holiday and with the purpose of then raising a second, similar sequence of events from the typical dry season baseline conditions.

The weather on 26 July, 1988 was satisfactory for the second venture. Overnight operations on 25/26 July with fixed-base units, aimed at triggering an anomalous west-to-east primary flow, appeared to have been successful. From 31 stories up on Singapore's south shore, a dark loom could be observed in the westerly sky. Later study of Changi radar plots verified that a moisture accretion had developed in the northwest quadrant from Singapore. The "gun" boat was taken due south from Changi on the morning of the 26th until we could turn due west and clear Singapore city in a long, straight "brewing" pass to the west.

This steady westerly thrust produced visible showers into the reservoir area on Singapore, and this was videotaped. By noon, the gunboat was southwest of Singapore city, with the small, spinning geometric device aboard etherically locked into the moisture mass northwest of Singapore, pulling the mass directly down on the gunboat and boosting its size and intensity. This was directly visible to us. Changi radar plots (next page) show the scenario precisely. At 11:00 AM the rain mass lies entirely to the north of the east-west Changi vector. Lightning is striking. By noon, the huge blob has oozed well south of the east-west Changi Vector, and is about to engulf western Singapore.

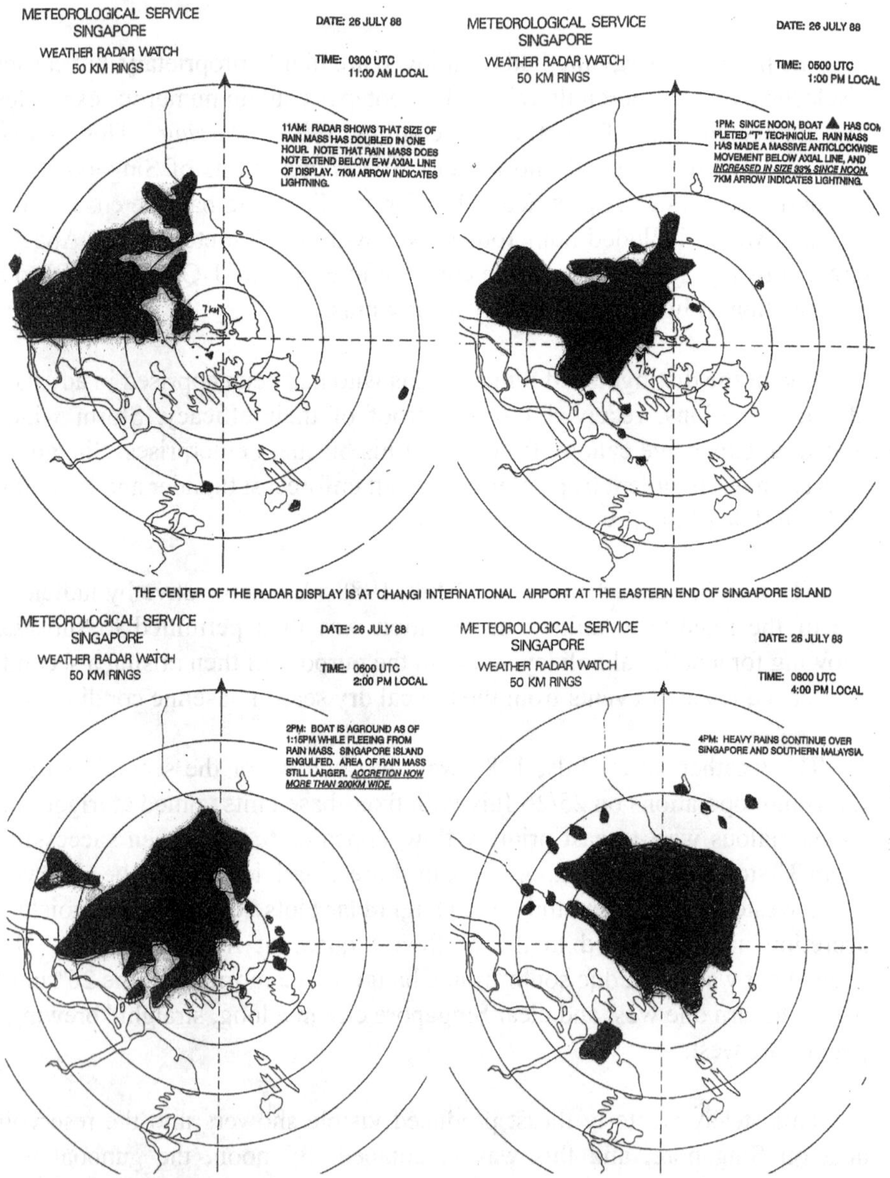

Our specialized "T" technique was implemented at 12:25. The 1:00 pm radar plot shows a doubling of the rain mass south of the east-west Changi vector. Lightning (arrow) continues. The dark, blue-black mass of moisture was now not only encroaching on Singapore city, but also was looming west and southwest of our boat, mountains high and threatening. The dramatic scene was videotaped and photographed, as the temperature began dropping steeply amid a rising wind.

By 2:00 pm, Singapore Island was engulfed, and so were we. Running like Dr. Frankenstein before the monster of our own creation, we ran the gunboat aground on a shoal. A deluge ensued, with zero visibility. Singapore Island remained under rain until after 4:00 pm. The official downtown rain gauge recorded 26.7 mm.

No further operations were conducted that day. Dry season conditions quickly re-established themselves, with a normal 27 July morning. Using a backup boat that was only marginally effective compared with the boat we had damaged, the "T" technique was employed again for two hours in the late afternoon of the 27th in sheltered waters off Loyang. A series of tremendous deluges ensued on the 28th, resulting in a further 59.9 mm into the downtown Singapore rain gauge, making a total of 114.8mm for the project period. Singapore reservoirs benefited with 68.3mm of rain in the same period. Thus ended TANGO, extending over 10 dry season days.

Vital functional information on primary energy in equatorial environments was gained from TANGO. The "T" technique is the most exciting innovation in this kind of engineering in many years, in the opinion of this writer. Because of its potential for destructive misuse, the basic procedure must remain highly proprietary.

Participating engineers in TANGO now have no doubt that the actual generation of low-pressure systems can be technically tackled and pushed through to success. By means of such engineered implosive vortices, comprehensive reversals of drought can be initiated, and probably sustained. Sagacious use of small, mobile, rotating geometric structures permits in the real world, and in the here-now, weather modification on a scale not feasible with chemical nucleation. The latter is the government-sustained method that absorbs over $100 million annually in the USA.

With a translator no larger than a human head, radar confirms that we raised about 30,000 square kilometers of rain mass over Singapore, Malaysia and northern Indonesia in the dry season. The engineering was conducted on the basis that the weather is a functional physical expression of the bio-geometric realities of the underlying ether. Weather engineering and control is thus approached fruitfully via the bio-geometric pathway, without chemicals or radiation. TANGO was another step along the bio-geometric path.

Update – 1991

We returned to Singapore in May 1991, creating more rain with a single "Skimmer" device we had put together. We were able to raise a rain mass over the Malacca Straight, south of Singapore, on 24 May 1991. Dense rain accretion was engineered from faair weather in about 30 minutes.

The "Skimmer" device consists of two concentric metal cones with an apex angle of 135 degrees. Cones are held at precise spacing within a PVC sleeve 8 inches in diameter. Rhythmic spinning motion and critical orientation of the device caused local blocking of powerful south-to-north flow of ether that prevails in May. Where the blockage is raised and temporarily held stationary by the rotating "Skimmer," very high etheric potential develops. This rise in potential is largely caused by the massive, northward-moving planetary flow piling into the engineered blockage from the south. High etheric potential strongly attracts water in the atmosphere, producing a kind of numbus. We witnessed the appearance of dark, horizontal striations that blocked the etheric flow from the south and spread laterally from the engineered etheric "dam."

The entire system eventually elevates to "lumination," or discharge potential, as we experienced here. Rain then descneds, often copiously in low tropical latitudes. Equatorial locales like Singapore, Malaysia and Indonesia make available decisive proof on the ether's physical existence, and accessibility to appropriate etheric engineering techniques. Engineered atmospheric effects are more rapidly and vividly obtained in tropical regions than in temperate, low humidity locales.

Adaptation of these techniques to airborne use has opened new vistas. Power and efficiency have been greatly increased, with the additional advantage of virtually unlimited mobility.

Chapter 19

OPERATION CLINCHER
Exclusive Background Report
Regional Smog Conquest
Southern California, 1990 Smog Season

by the staff of Etheric Rain Engineering Pte. Ltd.
in collaboration with Trevor James Constable

This specially prepared report is the first time that the historical work-up to 1990's Operation CLINCHER has been presented for the world public. While Etheric Rain Engineering Pte. Ltd. of Singapore no longer offers dedicated smog abatement contracts, the corporation may, in certain circumstances, make the technology available gratis to governments contracting for rain engineering operations. The extreme simplicity of the technology and the impossibility of protecting that technology via patents or other security measures preclude smog contracts as a practical matter.

ERE's predecessor groups were spectacularly successful against southern California smog in the 1987-90 period. Pre-notified etheric engineering operations against metropolitan smog began in 1987, with the pioneer project, Operation VICTOR. Prior to VICTOR, convincing evidence of the power of vertically-acting etheric vortices was obtained at sea, together with practical experience in their use.

Operation CLINCHER in 1990 was thus the result of lengthy, hands-on experience using etheric force in a vertical format. In no sense was CLINCHER a fluke, or a spurious, sudden entry into the smog scenario. The statistical record 1986-1990, graphed herewith (page 214), proves this decisively. The 14 stations used in CLINCHER could easily, today, be increased to 50 or 100 stations to wipe out this health-wrecking scourge. One percent of current smog budgets would finance such an effort. The truth is that political, bureaucratic and financial forces are *heavily invested in smog*, and exploit smog economically, financially and politically. This racket depends for its existence upon the lie that there is no effective technical answer to smog, only immensely expensive, marginally effective "projects" extending endlessly into the future. By contrast, presented here in their proper context, are the practical and theoretical precursors to Operation CLINCHER, the most successful of all ERE weather engineering operations since 1968. CLINCHER demonstrates an inexpensive, effective answer to this world's smog.

Development Background

Development in the mid-1980s, on a high seas ship, of "Flying H" type units occurred synchronously with operational divorcement from water grounding. Historical photographs of typical Flying H units reveal them as a side-by-side, H-type central mounting of a pair of resonant tubes on a common axle. Maritime testing of the Flying H began with manual rotation of the tubes, and provided proof of concept. Operations at sea established the need for a small but rugged a.c. gear motor to provide steady rotation in the frequent high winds of the ship's flying bridge. A powerful modern ship making 22 knots into a 25-knot headwind effectively has a full gale on its flying bridge. This was often the case on SS Maui.

Two examples of the Flying H

Once technical development departed from fixed, multiple-array, rack-type installations, and classical cloudbusters, into equipment incorporating rotating components on a motor-driven axle, many new approaches suggested themselves. Action is the crucible from which new designs emerge. The relative lightness and flexibility of Flying H units invited mobility. Having these rotating units mounted on a fast ship made it easy – and natural – for such equipment to describe typical spinning-wave or *kreiselwelle* waveforms in the ethers as the ship moved on its gyro-stabilized courses. Tentative new designs could be worked out, and then fabricated and tested on-site, in a pristine environment. Their practical value would be determined experimentally. The extensive operations that followed development of the Flying H soon showed that radically different equipment designs had moved our infant art into dynamic new domains of action and results. We had entered the revolutionary era of bio-geometric forms as a means of influencing etheric action.

The "Spider"

This new epoch of rotating components soon extended to the development of "Spider" type units, which were a logical outgrowth of the Flying H. The latter was essentially a horizontally influential device. "Spiders," by contrast, would work by generating vortices in the *vertical* plane. Original experiments along this line were carried out with vertical operations using the tested, tried and true "cone guns." These early, waterless derivatives of original cloudbuster technology married Golden Section metal cones to resonant PVC tube lengths. They were effective, but clumsy and awkward on shipboard in vertical use. Vessel motion added physical hazards. Simplification and stabilization were achieved by dispensing with the resonant tube section, and increasing the number of cones employed. This is how the classic "Spider" units developed. Their mode of operation was to direct the apices of the mechanically-rotated cones upward from the ship's deck toward the zenith. Any desired angle from the vertical could be engineered into the design. Infinite adjustment of these angles was provided in some designs. The cones could even be directed horizontally at their minimum adjustment. The operating theory was based on a known etheric property of a cone: that structure's ability to project a coherent beam of ether from its apex. Many people of even moderate sensitivity can detect this subtle radiation with the palm of their hand. The late, internationally noted dowsing master, Reverend Verne Cameron of California, was able to detect the coherent beam from a cone apex at a distance of several miles. The maximum range of the beam emitted from the cone apex has not yet been established. Creation and maintenance of the beam "out of nothing," by a geometric form alone, fundamentally challenges conventional concepts. No fuel or power source is required to establish and maintain the beam of etheric energy.

Typical Spider Unit

A Spider unit mounting two, four, or six cones in a circle around its perimeter, directs that number of subtle beams of etheric force upward, stirring the surrounding, moving ether to encourage vortex formation. The generation of such vortices is continuous as long as the Spider rotates. When the Spider is mounted on a moving ship, its action is greatly enhanced, on some courses, by the unit's own motion over the surface. Such a functioning Spider will trace the *kreiselwelle*, or spinning-waveform(s) in the ether. Objective effects are undeniable. Chief among these is the ability, in conditions of total overcast, to tear away this cloud cover above the ship, right through to blue sky overhead. The water vapor thus removed from the zenith, descends via vortical action to the local horizon in a dense, donut-type ring all around the vessel. This cloud ring and the opening torn through to blue sky above, move with the vessel over the ocean surface. These effects have been visually observed many times with radar confirmation, and an excellent daylight video example is found in the publicly available videotape, *Etheric Weather Engineering on the High Seas*.

Vortices At Work

In night replications of this technique, radar shows the moving vessel to be surrounded by a ring of highly active rain squalls, about 10 to 12 nautical miles in diameter. This entire rainy system, created by the Spider, moves synchronously across the ocean surface with the ship – sometimes for hours at a time. If the Spider is shut down, the ring of squalls soon dies. In normal operation, as that portion of the ring of squalls being left behind continually fades out astern of the vessel, fresh squalls materialize ahead to maintain the ring around the ship. The objective reality of these happenings is compelling, especially when they are observed for extended periods as stable phenomena. A video example of this type of activity, at night and on radar, is also found on the video *Etheric Weather Engineering on the High Seas*.

Implosive vortical action induced by the rotating cones entrains moisture as its contractive force thrusts down toward the ocean surface. This vortical action reinforces the natural, normal, nocturnal motion of the chemical ether back into the earth from the atmosphere – a daily occurrence that *does not produce this patterned activity without the Spider's intervention*. Hence the strong rain squalls. This astonishing activity is brought about with no chemicals of any kind, no electromagnetic radiation, and no electric power beyond that driving the small rotational motor. The overall happening is an evidential avalanche. The chemical ether "speaks" thus in its own language, a whole new fabric of natural law awaiting mankind's intelligent, cooperative touch.

During this break-in period with the Spider, in 1986-87, objective effects were produced on an even greater scale. The north Pacific "High" is more or less a meteorological fixture in that part of the ocean. Millions of square miles are usually involved in this blanket of relatively tranquil, high-pressure atmosphere. Judge of our astonishment when, during these vertically-active tests, we found that anomalous, sharp little "lows" suddenly started appearing near our positions on US Navy and US National Weather Service surface analysis fax maps. The indications were that these strange little systems had migrated from our ship's track, and were suddenly appearing in the middle of an almost inconceivably vast mass of high-pressure atmosphere that ruled a million square miles or more. These "implanted" mini-lows were appearing in sufficient strength and size to be observed and reported by surface ships and satellites. This practical experience of a subtle etheric force that *definitely worked physically*, led to theories as to how this action could be applied to some useful physical goal or purpose. A practical, beneficial value to humanity would be ideal. Further, wider use of etheric force for human advancement might well ensue.

Experience Teaches Best

Linkage already existed in nearly two decades of our shore operations to a modern problem with serious negative impact on humanity: smog. Our experience since 1968 in southern California had repeatedly demonstrated that *etheric rain engineering operations* definitely reduced air pollution levels in the region, without any intention to do so on our part. This smog reduction occurred adjunctively to etheric rain engineering operations, *whether or not rain was actually produced*. A "side-effect" is how physicians would describe this smog-reduction capability. Furthermore, Dr. James O. Woods and Trevor J. Constable had on several occasions, out of curiosity, directed large, water-powered rack units from Thousand Palms Oasis directly to the west through the Banning Pass and into the Los Angeles Basin. To our stunned astonishment, a heavy stench of smog and auto exhaust pervaded the pristine oasis area within a few minutes. Our action had physically moved smog constituents 50-60 miles eastward into the lower Mojave desert, almost instantly! There was thus a demonstrated technical link between etheric engineering and smog. This was verified by other, kindred experiences. With the heavy, cumbersome, water-powered equipment we relied on in those times, we could think out no feasible method for pursuing these findings. They were therefore tucked away in our memory banks for a later time. That time arrived with the Spider.

Maritime experience with the Spider had demonstrated convincingly that etheric vortices could be generated from a moving ship by these devices. We already knew there was a connection between the chemical ether and smog. Might it then be possible, perhaps, to generate implosive vortices in the Los Angeles Basin with Spiders operating at fixed bases? An implosive vortex entrains matter within its influence and drives such matter toward the vortex point. In a high seas format, the matter within its influence was atmospheric water vapor. That is how we had seen those rings of squalls surround the SS Maui and accompany the ship across the Pacific. While it was highly unlikely that similar operations from fixed shore bases in the semi-arid southern California area would produce squalls, the generation of milder vortices in such a dry climate might well have a beneficial and cleansing effect on the regional atmosphere around Los Angeles. Etheric vortices engineered there might entrain the airborne particulate matter that is integumented in smog and drive it back to the earth, producing significant air pollution reduction. This seemed worthy of further practical exploration.

Native Etheric Movement

In the fixed-based format on land, there would be no reinforcing influence from vessel velocity. We would be dependent entirely upon the *native* movement of the ether itself, passing through the vicinities of the emplaced Spiders. We already knew this motion to be continuous, based on the outline of the earth's ether economy provided by Dr. Guenther Wachsmuth in his monumental *Etheric Formative Forces in the Cosmos, Earth and Man*. Dr. Wilhelm Reich's brilliant pioneer work with cloudbusters in the 1950s left no doubt of the existence of a west-to-east flow of orgone energy (etheric force) in the northern temperate zone. In the eastern North Pacific, we had ourselves dammed up this west-east flow hundreds of times to engineer rain. We recorded the entire process unequivocally and often, on irrefutable time-lapse video. The existence and accessibility of this flow were thus as firmly established with us as our own heartbeats.

A rational expectation, based on our maritime experience, was that vortices stirred into this etheric flow by the action of the Spider(s) would move in strings "downstream" in the ether flow, away from Spider sites. They would possibly *influence implosively the physical atmosphere in those regions* – east of the Spider sites. In the fashion of vortex behavior, they could be expected to grow larger and weaker with distance, but would they travel far enough over dry land, and have sufficient influence to be physically and statistically evident? This was the chain of reasoning, theory and experience that brought us to the summer of 1987. Driven by the motto "only results count," we faced the Los Angeles July climate for our test of Spider-type units as smog-inhibiting devices.

Fortuitously, we found ourselves cheek-by-jowl with the most ambitious smog observation, monitoring and study project ever run by the California authorities, and probably unique in the entire world. This was a comprehensive program organized and funded by the California Air Resources Board in Sacramento, with a budget of $10 million. Dozens of smog scientists were brought to southern California from all over the U.S. and from other countries. These professionals were set up in numerous observation and collection bases at key sites in the four-county South Coast Air Quality Management District (AQMD) – the largest and filthiest such district in America.

Irrefutable Documentation

This was a stroke of good fortune for us. If our humble little project, essentially limited to the month of July, were to be successful, then such a detailed smog-monitoring project would show our influence unequivocally. We were elated to know that smog observations of unprecedented detail and complexity would be carried on by the State of California during VICTOR. Were we to fail then our impotence would show statistically and objectively, and justify abandonment of such smog operations in the future. Budgeted at a penny-pinching $4,000, our Operation VICTOR would utilize only three Spider-type stations for the month of July. The main installation was on Point Fermin, the southernmost tip of Los Angeles, on the Pacific Ocean. A second installation at Tustin, California, was close to the middle of the L.A. Basin, and the third was at Fort Zinderneuf in Desert Hot Springs on the edge of the Mojave Desert. These stations were equipped with hybrid versions of the evolving Spider, but all three units worked upward to initiate vortices in the ether.

Filed In Advance

VICTOR was filed in advance of commencement, with the National Oceanographic and Atmospheric Administration of the U.S. Federal government with same required form used by – and shown in this book – Projects PINCER II (page 172) and CLINCHER (page 203).

Incredible Events

The scenario that unfolded in southern California in July of 1987 strained credulity even among ourselves, poised as we were for positive results. Numerous weather and smog records were broken. The prototype Spider we had emplaced at Point Fermin produced a drastic reduction in temperatures for most of the month, and the signature of its influence was evident in satellite water vapor photos and 500-millibar charts broadcast on TV and eagerly videotaped by us. Cold air came rushing down right to Point Fermin from Canada, day after day. The atmosphere in the Los Angeles Basin was crystal clear. Video clips survive in our files of TV weathermen raving about the pristine visibility and clean air, and pointing excitedly to the January-in-July behavior of the cold Canadian air diving down to Point Fermin – our main operating base.

Scientists Go Home

Smog scientists in the State project were meanwhile struggling to find sufficient smog to justify their $10 million presence. There wasn't sufficient smog available for them to gather statistically valid or useful samples. This small army of professional people was accordingly sent home before the end of July. This humiliating abandonment of the smog project was also covered on TV. Compelling proof had been provided, documented by the State of California, that our generation of implosive etheric vortices would indeed reduce southern California smog. The month of July 1987 stood out for low smog readings within the 1987 season records, while *smog overall for the six months 1987 season took a striking 16.6 percent drop from 1986, down to a new all-time low!* A significant footnote, with VICTOR concluded in July, was smog returning with a vengeance in August. The banished smog scientists were then brought back for another study-try.

Victor for us was an encouraging major triumph. The AQMD smog bureaucracy had received copies of our advance Federal filing of Victor. They nevertheless palmed off on the media the excuse that this stupendous chain of clean air events was due to "freak meteorological conditions."

Early in 1988, Trevor Constable became associated with a young Singapore entrepreneur and international businessman, Mr. George K. C. Wuu, who became Chairman and CEO of ERE. A multimillionaire with a real sense of adventure, he began working with us to make etheric weather engineering commercially viable. As an almost ceaseless world traveler, Mr. Wuu had firsthand experience of the planetary scourge of smog, and envisioned the commercial potential of an effective and simple method of eradicating this curse. Here was truly a new way, inexpensive and effective. Further development was justified. He subsidized video production, equipment fabrication, transportation and other activity essential to keep the work moving. His intervention was both welcome and timely. The strain of 20 years was beginning to tell on the pioneers.

Upward Bulge

In the 1988 smog season, we stayed out of the Los Angeles smog scenario. Smog took a large upward seasonal jump, back to 77 Alert Days from the VICTOR all-time low of 66 days. This put smog back near to the same level it had reached in 1985 and 1986. Our signature was thus already in the smog statistics and graphs, even as we prepared to impress it there with even greater clarity over the next two seasons.

We continued working on Spider development in 1988-89. Designs were fabricated and tested until we had physically manageable units that were effective in generating etheric vortices. George Wuu had some beautiful examples professionally fabricated in Singapore, and installed them on his office roof. He took some dramatic, 360-degree "fisheye" photographs of the Singapore skyscape that bore graphic testimony to the tropical power of Spider equipment. We continued with rain engineering experiments in Hawaii, California, Singapore and on the high seas. With George Wuu's counsel and financial help, we decided in 1989 to run a pair of short projects against southern California smog that were within our capabilities and resources. The two operations would together cover three of the six months officially assigned as the smog "season." This would verify the effectiveness of the improved Spiders we were now using, in a challenging practical format. This was a significant step upward from the July-only Operation VICTOR, but there were human limitations on what we could do. Sufficient personnel were not then available to permit a full-season operation. We ran Operation BREAKTHROUGH in July, and Operation CHECKER in September and October of 1989.

The seasonal impact of these operations was spectacular. Smog Alerts dropped by a stunning 29.4 percent, all the way down to a seasonal total of 54 Alerts – an all-time record far below even the smashing VICTOR operation. This left little doubt that the increased efficiency and improved placement of our new units was responsible for this tremendous drop, since we had only used four stations. Our future in conquering smog looked rosy. Etheric technology could drive this scourge from the earth!

George Wuu decided on the basis of the 1989 success, that funding an enlarged, full-season smog operation in 1990 was a justified investment. A success to match the previous two might well put etheric weather engineering "over the top" as a commercial venture against smog. Plans were accordingly made to increase the number of Spider stations. More stations were essential. *As the number of Alerts was reduced significantly, reducing them still further would be difficult*. We therefore developed a simple, low profile, dependable and effective Mark II Spider that could be operated from any secure site that had a.c. power. The intention was to operate up to 14 stations if suitable sites could be found and commissioned. All this was arranged by the spring of 1990, and brings the story of CLINCHER to its chronological starting point. The account of CLINCHER that follows is an edited version of a Special Report originally published in the May-June 1991 issue of the *Journal of Borderland Research*, before becoming part of our ERE website. Compiled by Thomas Brown, then-director of Borderland Sciences Foundation and now Etheric Rain Engineering's distinguished webmaster, this 1991 Special Report excluded, for space restrictions necessary at that time, the enlightening technical details covered in this Background Report/Preface. With the information provided in both articles, the reader can grasp that CLINCHER was no fluke or freak occurrence. Thousands of hours of dedicated labor lay behind the smashing triumph that CLINCHER became, and yet CLINCHER remains, as of this writing, the *air pollution technology that nobody wants*.

Chapter 20

OPERATION CLINCHER
Seasonal Smog Reduction, Southern California, 1990

A Special Report compiled and edited for the *Journal of Borderland Research* by Thomas Brown, Director, Borderland Sciences Research Foundation

Operation CLINCHER was the climactic operation in a series of four etheric weather engineering projects carried out by Trevor Constable and his group. All four projects were aimed at air pollution reduction in Southern California. As required by law, all the projects were filed in advance of commencement with the National Oceanographic and Atmospheric Administration, Rockville, Maryland.

CLINCHER developed out of Trevor's original operational experience with vertical ether currents. Simple geometric apparatus was designed to generate etheric vortices, which can subsequently influence the atmosphere. This phase of weather engineering began early in 1987 on the high seas, aboard a large, fast-moving, ocean-going vessel.

Promising mobile experience justified an exploratory fixed-base operation in southern California in July 1987, code-named VICTOR. This operation was a striking success against smog. Significant ozone reduction resulted together with the cleanest air seen in the region since the start of the smog records. There were unforeseen consequences, for which there had been no malefic intent, but which provided forceful, objective evidence of VICTOR'S efficacy.

A $10 million smog study project mounted in the summer of 1987 by the California State Air Resources Board was nullified by the simultaneous VICTOR operations. The scientists that had come to southern California for the study, from all over the U.S. and other countries, were sent home because there wasn't enough smog for them to gather statistically valid samples. The most ambitious smog program ever launched was an ignominious failure. An effective anti-smog modality had come on the scene from "left field," and history was in the making.

Trevor Constable's ATMOS group stood down from all southern California operations in 1988, during which time Project TANGO was undertaken in Singapore. (This operation is detailed elsewhere in this book.) California smog levels were permitted to "normalize" in that year, i.e. to develop without any countering influence from etheric weather engineering. This stand-down was part of a necessary rebuttal to the official assessment of 1987's dramatically low smog as being due to a "meteorological fluke."

1988 returned the southern California smog season to 1986 levels with 77 Alert Days – up almost 17 percent from 1987.

In 1989, the Constable group returned to southern California. BREAKTHROUGH was mounted in July, and CHECKER followed in September and October. This totaled three months of anti-smog weather engineering operations during the six months of the 1989 season. A smog season runs from 1 May through 31 October annually. George K. C. Wuu, a successful young Singapore entrepreneur had now begun subsidizing etheric weather engineering.

In 1989, operations provided a sparkling reprise of the 1987 VICTOR scenario. Smog was yet again reduced to all-time record low levels. Thus, in two separate years, two all-time low smog levels were recorded with a telltale peaking to "normal" between them, when Constable's ATMOS team did not participate.

This operational sequencing provided typical ON/OFF statistical evidence. Official seasonal smog status in Los Angeles is assessed via the number of Alert-Days in a season. An Alert-Day is one on which there is a First Stage Smog Alert anywhere in the South Coast Air Quality Management District, hereinafter the AQMD. A First Stage Smog Alert is called whenever ozone at any monitoring station reaches .20 parts per million. There are 37 official monitoring stations in the AQMD, which covers four large counties, including San Bernardino County, the largest county in America.

1986 Atmos Out

The only ATMOS operation in 1986 was Operation PINCER II, in July only, detailed earlier in this book. This was exclusively a rain operation, utilizing a different technology than is required for smog operations. 1986 was a normal smog season. 79 Alert-Days.

1987 Atmos In

Operation VICTOR Phase 1. 1 July through August 9. Phase 2, 1-30 September. Produced the cleanest southern California air in 40 years. Lowest number of Alert-Days ever recorded (66). A 16 percent drop in Alert-Days from 1986. Termed a meteorological fluke by smog officialdom. 66 Alert-Days.

1988 Atmos Out

No smog operation. Smog returns to 1986 levels; up 16 percent from 1987. 77 Alert-Days.

1989 Atmos In

Operation BREAKTHROUGH in July. Operation CHECKER in September and October. Total engagement time: 3 months, or half the smog season. New all-time low number of Smog-Alert Days: 54, a 29 percent drop from 1988. 54 Alert-Days.

The statistical correlation between ATMOS operations and smog levels thus already showed a high degree of probability. CLINCHER was designed and mounted, in the words of TJC himself, "to raise that probability to a level where would-be skeptics would appear ridiculous."

Operating Criteria

Criteria determining CLINCHER'S pre-filed operational goals were a follows:

1. The margin under 1989 had to be significant, substantial and noteworthy. No marginal reduction that could be washed out by statistical manipulation would suffice. Unannounced changes in AQMD statistical bases are known to have occurred previously.

2. The combined Alert-Days reduction of the two successive ATMOS years, 1989 and 1990, had to be unprecedented in the history of the records. This reduction had to exceed by an inarguable margin, any other two successive non-ATMOS years in the records.

3. The reduction in Alert-Days from 1989 to 1990 should be larger than any previous year-to-year attrition in non-ATMOS years when the AQMD was performing entirely on its own.

4. The CLINCHER reduction had to be feasible with available ATMOS resources. A 50 percent reduction was now technologically feasible, but not financially possible with available funding and personnel.

5. The final result had to CLINCH the question of etheric weather engineering's influence over smog. Hence, CLINCHER.

A 20 percent reduction in Alert-Days by CLINCHER was therefore chosen as best meeting these criteria.

New Operational Factors

In the 1989 CLINCHER operation, success probability was enhanced by three factors not previously present:

1. At least SIX operating sites would be used, 12–14 if sites were available. A maximum of four sites had been used previously.

2. Operations would be conducted for the *full season* for the first time, at least doubling the project's leverage over smog.

3. Advances in effectiveness of equipment had been achieved since VICTOR in 1987. A major technical advance had been made after the 1989 season.

These factors united behind Trevor Constable publicly setting a 20 percent reduction (full season) as the project goal for CLINCHER. This projected reduction was posted in the Federal filing for CLINCHER made with NOAA 6 April 1990. Furthermore, record seasonal reduction of regional smog was stated on the Federal Initial Report as the purpose of the activity.

There were no ifs, no waffling, no hedges. *Smog was being challenged head-on.*

This announcement of a further drastic smog reduction, *below the all-time record*, prior to project commencement, was an audacious commitment. Constable further made it public internationally by announcing it on Radio Free America. Through the RFA shortwave outlet, the announcement reached millions. The *Journal of Borderland Research* (May-June 1990) also published the announcement in advance. Aware that he was stepping on the toes of some very powerful people profiting from smog, Constable avoided all local and regional publicity to minimize any "spoiler" activity.

No orthodox scientific body, no responsible bureaucracy or bureaucrat has ever dared give the public this kind of unequivocal commitment for effective air pollution reduction. The southern California smog bureaucracy had been plainly flabbergasted by the 29 percent Alert-Days reduction in 1989. The

AQMD had no expectation that such a reduction would be repeated in 1990. Dr. James Lents, executive director of the AQMD, admitted at the 1990 mid-season AQMD press conference that smog had already been far lower than expected – even allowing for favorable weather.

Dr. Lents, together with AQMD Chief Scientist Alan Lloyd, and AQMD Chief Meteorologist Joe Cassmassi, had been sent copies of the CLINCHER Federal Filing before the season started. No such corporate correspondence is ever answered of acknowledged. Certified Mail was used for these advices to preclude any later denials of receipt by officials.

The wherewithal does not exist within orthodoxy for a 15-20 percent regional reduction of smog in one season, *let alone TWO such reductions in successive seasons*. On the contrary, orthodox scientific opinion and conventional technical inadequacy have condemned the people of southern California to ponying up over $10 billion by the year 2000 to "*fight smog*." (Note: This has now eventuated.)

When Trevor Constable made his prediction of a 20 percent smog reduction in 1990, he really stuck his neck out. Plenty of know-it-alls got ready to throw a horse collar on his neck.

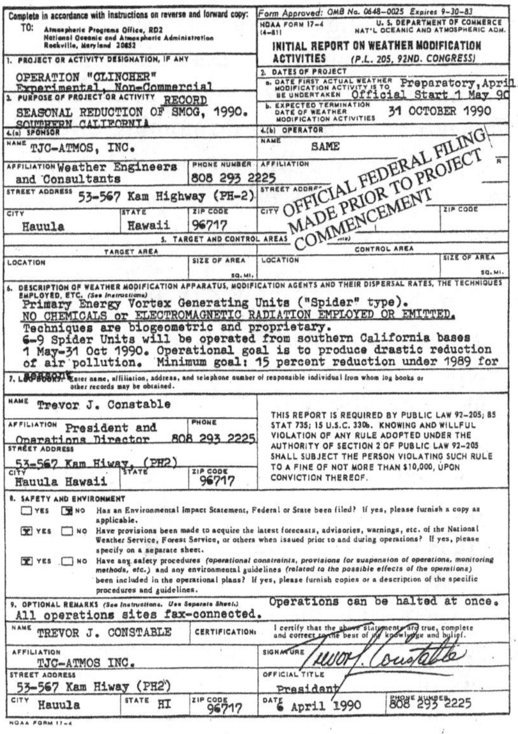
Complete in accordance with instructions on reverse and forward copy:
TO: Atmospheric Programs Office, RD2
National Oceanic and Atmospheric Administration
Rockville, Maryland 20852

Form Approved: OMB No. 0648-0025 Expires 9-30-83
NOAA FORM 17-4 (4-81)
U. S. DEPARTMENT OF COMMERCE
NAT'L OCEANIC AND ATMOSPHERIC ADM.
INITIAL REPORT ON WEATHER MODIFICATION ACTIVITIES (P.L. 205, 92ND. CONGRESS)

1. PROJECT OR ACTIVITY DESIGNATION, IF ANY
OPERATION "CLINCHER"
Experimental, Non-Commercial

2. DATES OF PROJECT
a. DATE FIRST ACTUAL WEATHER MODIFICATION ACTIVITY IS TO BE UNDERTAKEN: Preparatory, April. Official Start 1 May 90
b. EXPECTED TERMINATION DATE OF WEATHER MODIFICATION ACTIVITIES: 31 OCTOBER 1990

3. PURPOSE OF PROJECT OR ACTIVITY
RECORD
SEASONAL REDUCTION OF SMOG, 1990.
SOUTHERN CALIFORNIA

4.(a) SPONSOR
NAME: TJC-ATMOS, INC.
AFFILIATION: Weather Engineers and Consultants
PHONE NUMBER: 808 293 2225
STREET ADDRESS: 53-567 Kam Highway (PH-2)
CITY: Hauula
STATE: Hawaii
ZIP CODE: 96717

4.(b) OPERATOR
NAME: SAME
AFFILIATION
STREET ADDRESS
CITY
ZIP CODE

OFFICIAL FEDERAL FILING MADE PRIOR TO PROJECT COMMENCEMENT

5. TARGET AND CONTROL AREAS
TARGET AREA
LOCATION
SIZE OF AREA SQ. MI.
CONTROL AREA
LOCATION
SIZE OF AREA SQ. MI.

6. DESCRIPTION OF WEATHER MODIFICATION APPARATUS, MODIFICATION AGENTS AND THEIR DISPERSAL RATES, THE TECHNIQUES EMPLOYED, ETC. (See Instructions)
Primary Energy Vortex Generating Units ("Spider" type).
NO CHEMICALS or ELECTROMAGNETIC RADIATION EMPLOYED OR EMITTED.
Techniques are biogeometric and proprietary.
6-9 Spider Units will be operated from southern California bases 1 May-31 Oct 1990. Operational goal is to produce drastic reduction of air pollution. Minimum goal: 15 percent reduction under 1989 for season.

7. LOG BOOKS: Enter name, affiliation, address, and telephone number of responsible individual from whom log books or other records may be obtained.
NAME: Trevor J. Constable
AFFILIATION: President and Operations Director
PHONE: 808 293 2225
STREET ADDRESS: 53-567 Kam Hiway, (PH2)
CITY: Hauula Hawaii
STATE
ZIP CODE: 96717

THIS REPORT IS REQUIRED BY PUBLIC LAW 92-205; 85 STAT 735; 15 U.S.C. 330b. KNOWING AND WILLFUL VIOLATION OF ANY RULE ADOPTED UNDER THE AUTHORITY OF SECTION 2 OF PUBLIC LAW 92-205 SHALL SUBJECT THE PERSON VIOLATING SUCH RULE TO A FINE OF NOT MORE THAN $10,000, UPON CONVICTION THEREOF.

8. SAFETY AND ENVIRONMENT
[ ] YES [X] NO Has an Environmental Impact Statement, Federal or State been filed? If yes, please furnish a copy as applicable.
[X] YES [ ] NO Have provisions been made to acquire the latest forecasts, advisories, warnings, etc. of the National Weather Service, Forest Service, or others when issued prior to and during operations? If yes, please specify on a separate sheet.
[X] YES [ ] NO Have any safety procedures *(operational constraints, provisions for suspension of operations, monitoring methods, etc.)* and any environmental guidelines *(related to the possible effects of the operations)* been included in the operational plans? If yes, please furnish copies or a description of the specific procedures and guidelines.

9. OPTIONAL REMARKS *(See Instructions. Use Separate Sheet.)*
Operations can be halted at once.
All operations sites fax-connected.

NAME: TREVOR J. CONSTABLE
CERTIFICATION: I certify that the above statements are true, complete and correct to the best of my knowledge and belief.
AFFILIATION: TJC-ATMOS INC.
SIGNATURE: Trevor J. Constable
STREET ADDRESS: 53-567 Kam Hiway (PH2)
OFFICIAL TITLE: President
CITY: Hauula
STATE: HI
ZIP CODE: 96717
DATE: 6 April 1990
PHONE NUMBER: 808 293 2225

NOAA FORM 17-4

Says Constable, "History confirms unerringly that orthodoxy's reaction to truly radical developments is to *bash them into the ground*. This is usually achieved when such advances are germinal, and their acceptance turns upon minor, hair-splitting matters such a miniscule differences in meter readings or temperatures and so forth. The inventor and the idea can then be crushed by dishonesty, deceit, stupidity, incomprehension, neurotic evasion, or all five at once."

Some readers will recall Einstein's scuttling of Dr. Wilhelm Reich's orgone accumulator. Einstein, the man who made the ether superfluous via mathematical gambits, could not accept that Reich could concentrate the ether in a layered box, and create an irrefutable thermic differential with the arrangement. The differential had to be verbalized away – evaded. There are numerous other kindred examples and the names of Nikola Tesla, Dr. Albert Abrams and Dr. Ruth Drown come immediately to mind.

Trevor Constable pungently expressed his standpoint on the institutionalized evasion that blocks what is radically new. "The nazified treatment orthodoxy gives people like the late Dr. Reich, for example, jailing him and burning his books and papers, is a disgrace to science. Jail for serving and helping mankind? I don't like and I don't trust such barbarians, and they run the whole bloody show in science and the government. *Smog has them flummoxed.* They can do nothing effective. They 'study' smog, facilitate the profit plunder from smog, and try to block and sequester the effective remedy. I do not seek their approval because it is irrational to do so in the circumstances…"

In the case of CLINCHER, nothing miniscule or marginal, or subject to nitpicking argument, was involved. There would be no controversy over fractions of a Fahrenheit degree. CLINCHER was on a vast, mind-boggling scale. *Air pollution in the largest, worst afflicted region in America, four huge counties in extent, was to come down seasonally by one-fifth.* Could such a colossal reduction actually be achieved two years in a row? All this by "etheric engineering" when orthodoxy remains compulsively convinced that there is no ether? Could it be done?

The men in the ATMOS group were in no doubt. According to veteran aide Irv Trent, who died at the age of 86 in 2001: "TJC devoted over 20 years of his life, and a professionally-earned fortune, to bring everything together in CLINCHER. His operational experience is unrivaled, and his team loves him. For CLINCHER, he called in every marker. He put his considerable international reputation on the line. I knew CLINCHER would be a smashing success." The scholarly former aerospace technician added a prescient estimate. "This time," he said, "the etheric revolution is going to bash orthodoxy into the ground. Einstein's etherless universe is going into the ash can, where it belongs."

Problems remained. In prior smog seasons, Riverside and San Bernardino had always been difficult areas. This is because smog migrates from the central and western Los Angeles Basin eastward into these communities. Mountains block further eastward drift. Smog Alerts result. The most powerful CLINCHER installation, a Mark 7 Spider, was sited at 3400 feet in Banning, and would definitely affect San Bernardino, lying due west magnetic. A potential hazard was that part of San Bernardino's smog might shunt southward into Riverside. Spider units in Riverside were crucially needed.

TJC was unable to obtain any Riverside bases. The base existing at Perris was a little too distant, on its own, to keep Riverside Alerts in check. Nobody would help. A spider occupies about the same space as a kitchen chair, and uses a small AC motor to power its rotation. With no chemicals and no electromagnetic radiation, the device is virtually noiseless and environmentally pure. Nevertheless, there were no Riverside takers to appeals via the Chamber of Commerce and civic clubs.

Desperate to obtain a foothold in Riverside, TJC wrote a corporate appeal to Riverside supervisor A. Norton Younglove, who also served as Chairman of the Board of the AQMD. Younglove was asked if he could help the ATMOS group obtain a couple of operating sites in Riverside. He was advised that this could help diminish a curse on his constituents always hard hit by summer smog. For a veteran politician like Younglove, it was a minor request of potentially great benefit to his electors – at no cost. He could have helped mightily with a couple of telephone calls.

Younglove completely ignored this request. He never replied to a polite suggestion that he view an ATMOS corporate videotape providing full background to the CLINCHER operation. Younglove was not alone. The California State Air Resources Board declined a request to give Constable a 10-minute hearing at one of their meetings to explain personally the nature of the unorthodox etheric approach. This offer was made by the president of a private corporation that had invested more than half a million dollars in original weather research, now sharply germane to California's major pollution problem. That same man, Trevor James Constable, through his widely translated aviation histories and biographies, enjoyed an international literary standing and respect for his integrity. Corporate videotapes sent to the Air Resources Board and to its scientific division evoked no response, not even acknowledgement of receipt.

These attitudes and reactions of officialdom show that the California public was, and still is, *ill-served by those entrusted with the reduction of air pollution*. TJC's trusty aide, the late Irv Trent, was a lifelong deep student of politics, and said, "Smog has actually become an industry in its own right. The truth is that nobody in the seats of power wants it sliced down in 25 percent annual chunks, the way TJC demonstrably can do it right now. He could cut it 50 percent more with new finds from CLINCHER and 40 to 60 small vortex generators." Surely this was what California wanted in 1990 and still needs today... billions later. Trent remained, until his death, without illusions on that score. "Smog vanishing on that scale," he said, "is the derailment of a gigantic gravy train. Businessmen programming profits from the billions in coerced capital investment will lose their shirts. Politicians see rapid conquest of smog as the end of smog graft. Bureaucrats see massive smog reduction as termination of their empire building. The public doesn't have a chance against this crooked combination with its criminal selfishness and vested interest in smog."

CLINCHER began with six Spiders – basic generators of etheric vortices. The principle is direct, and is based upon the existence and technical accessibility of the ether. That element of the ether with which Constable deals, he describes as "a physical natural force of extreme subtlety but tremendous power that is geometrically accessible." Constable is aware that this is indigestible by the established order in physics. "The directly visible control of local weather on the high seas that I have repeatedly demonstrated publicly on time-lapse videotape – at horizon distance – is something not feasible by any other method," he says. "Victims of parrot education squawk that there is 'no ether, no ether, no ether.' I have proved via real-world results, rather than talk or theory, that the ether is an objective reality that is technologically usable in countering air pollution. Theories are cheap and plentiful. Theories are ineffective against smog and mankind suffocates. Only results count. Etheric technology gets results."

Constable's basic approach is that the ether underlies the atmosphere in all its workings. He sees the atmosphere as the slave of the ether. Vortices geometrically induced in the ether – at the primary or etheric level – will translate into the atmosphere and develop into vortex strings from the original disturbance. The largest of these vortex chains are sometimes seen on official Surface Analysis weather maps as low-pressure systems in strings – two, three, and sometimes four in a row. Constable says that during May and June 1990, such strange strings of lows extended in straight lines from southern California all the way to northern British Columbia and southern Alaska, "carried on and in the south-to-north summer flow of etheric force."

ETHERIC VORTICES IN ACTION

The U.S. Navy Western Oceanography Center at Pearl Harbor, Hawaii, originated this radiofacsimile surface analysis weather map. The Valid Time is midnight, Greenwich time, 10 May 1990. After start-up of Operation Clincher in Los Angeles on 1 May 1990, TJC noted that a string of low pressure barometric systems formed (Margin designation A,B,C,D) from southern California (A) to Alaska (D). Similar atmospheric vortices continued to appear in strings for the next two months in this anomalous fashion. TJC interprets their appearance, and persistence, as due to vortical activity induced by CLINCHER apparatus operating 24 hours a day in Los Angeles. Such vortical activity in the main south-north flow of etheric force in spring and summer (northern hemisphere) behaves similarly to vortices in other media. Etheric vortices translate into the atmosphere eventually, as indicated here by this 2500 mile-long vortex string. Inducing such implosive vortices locally is the key to mastering smog, according to TJC. The real-world consequence of the persistently appearing vortex strings above, was the wettest northern California spring in 60 years, and obliteration of drought in the Pacific northwest. Also obliterated were doomsday, computer-born predictions of drought disaster in the Pacific northwest. The on-site rain gauge at BSRF headquarters in coastal northern California measured 19" in May and early June 1990, during predictions of drought and an early fire season.

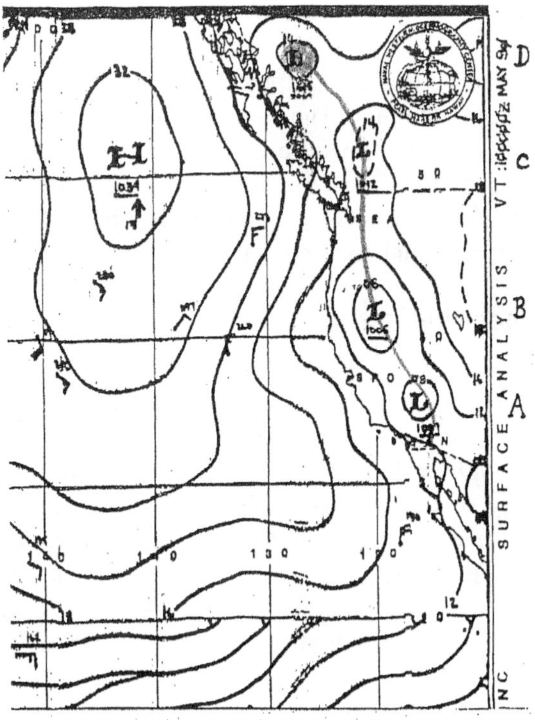

Thousands of lesser and miniature implosive vortices result locally as well, initiated by and migrating from each Spider unit – in accordance with seasonal, lunar and diurnal etheric flow laws. "These vortices spread out in the vast rivers of etheric force that flow through the physical world unperceived and unsuspected by most humans," says Constable. "These vortices probably entrain subtle particulate matter, driving it to the ground via the vortex points." This engineered implosive activity in the ether is his key to countering smog, by dynamizing torpid atmosphere over hundreds of square miles.

Los Angeles Basin topography provides a natural trap for smog build-up through physical confinement and inadequate atmospheric movement. Orthodoxy accepts that. The deceptively simple Spiders act against atmospheric stasis, and help re-establish etheric balance. Spiders work exclusively on the etheric continuum, even though they are themselves physical assemblages. The implosive vortical activity initiated by the Spiders also *strongly inhibits ozone*. Spiders, furthermore, help disperse the notorious inversion layer over the Los Angeles Basin, an atmospheric "lid" that holds smog against the earth, confined by the surrounding mountains.

This unique topography causes the smog build-up for which Los Angeles is justly infamous. The sun adds photochemical reactions to the lethal brew. Windy, vibrant days are common in southern California when the Spiders are operating. Constable firmly believes that it is possible to eliminate the inversion layer over Los Angeles via etheric engineering, returning the region to clean, primal conditions by dispersing all stasis.

Clincher Starts

On 1 May 1990, CLINCHER opened with Spiders in six southern California communities. Within two weeks, eleven Spiders were operating. Two fabulous months ensued, as Alerts were decisively reduced even under the 1989 level. From 1 May until the end of June, there had been only 11 Alert-Days in the entire AQMD. Visibilities were phenomenal, and residents were treated to stunning views of the mountains around the Los Angeles Basin, lost to sight for years, and now visible for weeks on end. Nothing like it had been seen in years. From the Palos Verdes Hills, at the southern tip of Los Angeles County, 40-mile view vistas of the coast, and shimmering landward seas of lights went on week after week. The "beautiful weather" was a happy public topic, and the entire region reveled in benign conditions. The AQMD remained strangely quiet. The smog bureaucracy was desperately grubbing for a $20 million boost in its annual budget as the region sailed on in the cleanest air on record. The AQMD got its budget boost to $100 million annually. CLINCHER'S entire budget was $35,000.

As the season reached the end of July 1990, the effects of the Spider installations were significant in the mid-year statistics. The unit in Altadena-Pasadena exerted wide influence beyond its own locality. Pasadena itself was reduced at least 50 percent in Alerts under 1989. The Spider site at Reseda resulted in zero Alerts for the entire season – a shutout. San Bernardino Alerts were drastically reduced, and the troublesome Orange County area was now among the best. Traditionally smoggy spots along the San Gabriel mountains – a barrier at the north end of the L.A. Basin – all showed large reductions in Alerts. Even Glendora, the nation's smoggiest community, was reduced 24 percent below 1989 Alerts.

By August, the 1990 smog season was being identified on TV and in newspapers as the cleanest ever. This startling fact of regional life was suppressed by the *Los Angeles Times*, the major regional medium. L.A.

Times editors killed reports of the AQMD mid-season press conference. The newspaper had been made aware of CLINCHER, although no publicity was wanted from any local or regional medium. This public silence was considered essential by TJC as prophylaxis against the paralyzing nuisance lawsuits that characterize and blight American life.

Mid-season's *bete noir* was Riverside, which was running ahead of its 1989 Alert rate, even as adjacent San Bernardino's Alerts had been reduced by more than 60 percent. This was serious. Some of San Bernardino's smog was being shunted into Riverside where no Spider sites had been obtainable, and where local Supervisor and AQMD Board Chairman Younglove had declined assistance.

An internationally famous gentleman now made a decisive entry into the CLINCHER drama. TJC was taken to Riverside finally to meet personally General Curtis LeMay, USAF Retired, former Chief of Staff of the USAF, creator of the Strategic Air Command, and a longtime fan of Constable's aviation histories. General LeMay had kindly provided, just a few months previously, a jacket blurb to Constable's latest biography, *Fighter General: The Life of Adolf Galland*, first published in June of 1990. In addition to General Galland of Germany, TJC and General LeMay had numerous mutual friends among aviation's luminaries. In 1968, TJC had worked for the Wallace-LeMay presidential campaign, a seemingly vain labor that was about to come full circle.

Immediate accords developed between Trevor Constable and General LeMay and his wife, Helen. A notorious tinkerer and "can do" guy, LeMay immediately wanted to view the ATMOS videotape. When LeMay met TJC on 11 August, after viewing that videotape several times in private and on his own, the General had his right forefinger raised admonishingly, his pipe in the other hand. "You never convinced me one bit with that videotape. Not one bit. *But I want one of those damned things in my back yard.*"

TJC had one of his Spider units in his car, certain that he would get the General's help. He pulled it out, and the former Chief of Staff of the United States Air Force *produced a wrench and helped him install it.* General LeMay's aides, one of them from Supervisor Younglove's office, rapidly provided two additional bases in Riverside. One base was downtown, the other in strategic Norco. CLINCHER was now operating from 14 bases. A dramatic turnaround ensued in Riverside smog.

In the 1 May to 11 August period, Riverside experienced one Smog Alert for each 9 days. After the intervention of General LeMay, from 11 August through 31 October, Riverside experienced one Smog Alert for each 20 days. That brought Riverside's season total in with 15 Alerts for the 1990 season, against 17 in 1989, an 11.7 percent reduction. Without General LeMay, Riverside would have been a nasty blot on CLINCHER'S statistical triumph. The promised 20 percent reduction in Alert-Days would probably have become marginal. The action of a single, outstanding individual had made a decisive difference.

The time nexus of General LeMay's involvement also demonstrated objectively what Spiders could do in a heavily smogged area, *producing more than 50 percent reduction in Alerts after installation.* This "project within a project" provided valuable, documented evidence for the records.

Said Trevor Constable of General LeMay: "His last involvement on this earth was directly to benefit every one of his fellow citizens in Riverside, right down to babies. He performed a wonderful, life-giving deed. He was vital to our victory. General LeMay remains the hero of CLINCHER, and a hero to me and my men. I took his passing deeply to heart."

General LeMay, at 83, was full of ideas and plans to use Spiders for clearing fog at USAF bases afflicted with this problem. He had seen the smog-veiled mountains east of his home come into view after the Spider was installed in his patio, and stay visible. He was proceeding to get USAF interest when he died suddenly, early in October 1990, a shattering loss to the USA and to the CLINCHER crew.

When taking what was to be his final leave of Trevor Constable, his words were typical LeMay and also prophetic. "It doesn't matter if we don't know ALL about this ether thing right now. We'll find that out. You're doing what has to be done. Just go right at them and *let'em have it.*" This legendary commander served as an inspiration to the CLINCHER TEAM. In seeking to serve mankind through many years, his involvement gave them a hand, a morale boost and vital assistance. Constable found the General unforgettable for one trait: "Curtis LeMay is one of the few men I have known in my lifetime who knew how to listen. He contained his own reactions completely, which few men can do, while you were speaking. He was totally locked onto you. When you were through, he had a basic comprehension of what you had said, no matter how novel or off-the-track. That's rare in this world."

By the end of the smog season on 31 October 1990, CLINCHER was operating from 14 bases and had made history. Southern California had experienced its all-time record low smog season exactly as laid out in the Federal filing in April. The L.A. Times ran a story covering the smog season with a statistical tabulation that appears herewith. CLINCHER was ignored by the smog bureaucracy, which had been kept apprised of all CLINCHER developments since April. The only mention in print came from the "alternative" press, the *Journal of Borderland Research*, and Dr. Antony Sutton's *Future Technology Intelligence Report*.

Only Results Count

Here are the major results of Operation CLINCHER:

1. REDUCTION in seasonal Alert-Days by 24 percent below 1989. This reduction exceeded by 4 percent to CLINCHER target announced in April. (The Alert-Day is essentially a statistical device for assessing seasonal smog levels, and is not an objective quantification.)

2. REDUCTION IN DURATION of First Stage Alerts by 60 percent under 1989. 1990 Alerts recorded lower levels of ozone, ant the Alerts *lasted less than half as long as in 1989*. This was a public service of the highest order at no cost to the public.

3. ACTUAL REDUCTION IN THE NUMBER OF ALERTS under 1989 was staggering. In various individual smog-monitoring areas of southern California, the drop in the number of Alerts was far more spectacular than the reduction of abstract Alert-Days used for assessing the entire region. Actual ALERT REDUCTIONS were, typically:

San Bernardino DOWN 68 percent.
Pasadena DOWN 58 percent.
Norco DOWN 100 percent (ZERO ALERTS)
Reseda DOWN 100 percent (ZERO ALERTS)
Anaheim DOWN 100 percent (ZERO ALERTS)
El Toro DOWN 100 percent (ZERO ALERTS)
West L.A. DOWN 100 percent (ZERO ALERTS)
Glendora DOWN 24 percent
Azusa DOWN 56 percent
Upland DOWN 42 percent
Redlands DOWN 41 percent
Riverside DOWN 11 percent

The only blot on CLINCHER'S otherwise perfect record was 2 Alerts in Downtown L.A. versus 1 Alert in 1989. A 100 percent rise.

4. HEALTH BENEFITS. Dr. Robert Phalen, who directed the Air Pollution Health Effects Laboratory at the University of California, Irvine, opined in the *Los Angeles Times* that 1990's air quality improvements were significant enough to *benefit the health of everyone living in the Los Angeles Basin.* This universal health benefit to southern Californians is of inestimable public value, and the jewel in CLINCHER'S crown.

Operation CLINCHER

1990 Smog Season Had Cleanest Air on Record

SMOG SEASON

The 1990 smog season, which started May 1 and ended Wednesday, was the cleanest on record in the Los Angeles Basin, which includes Orange, Los Angeles, San Bernardino and Riverside counties. Air is measured daily in 37 cities. This chart reports the number of smog alert days in selected cities. Officials say recent rules forcing cutbacks in emissions from industry and autos get most of the credit for the 24% basinwide improvement compared with 1989.

■ SMOG ALERT DAYS

City	1990	1989	1988	1987	1986	1985	1984
Glendora	28	37	54	51	70	68	67
Azusa	13	30	33	26	45	48	55
Pasadena	7	17	18	15	33	41	49
Upland	11	19	25	23	38	39	41
San Bernardino	7	22	31	27	41	30	36
Redlands	10	17	25	26	22	31	26
Norco	0	3	7	9	12	20	19
La Habra	4	5	2	6	8	13	14
Downtown L.A.	2	1	2	2	8	9	8
Reseda	0	5	4	2	5	9	6
West L.A.	0	1	2	1	1	4	5
Anaheim	0	4	1	3	1	11	5
El Toro	0	2	2	0	1	7	3
Costa Mesa	0	0	0	0	0	1	1
Long Beach	0	0	0	0	0	1	1
Los Alamitos	0	0	1	0	0	0	0
Total in basin*:	41	54	77	66	79	83	97

*The basin total is the number of days when at least one monitoring station within the four counties reached the smog-alert limit for ozone, which is .20 parts per million.

Source: South Coast Air Quality Management District

LOS ANGELES TIMES, 1 NOV. 1990

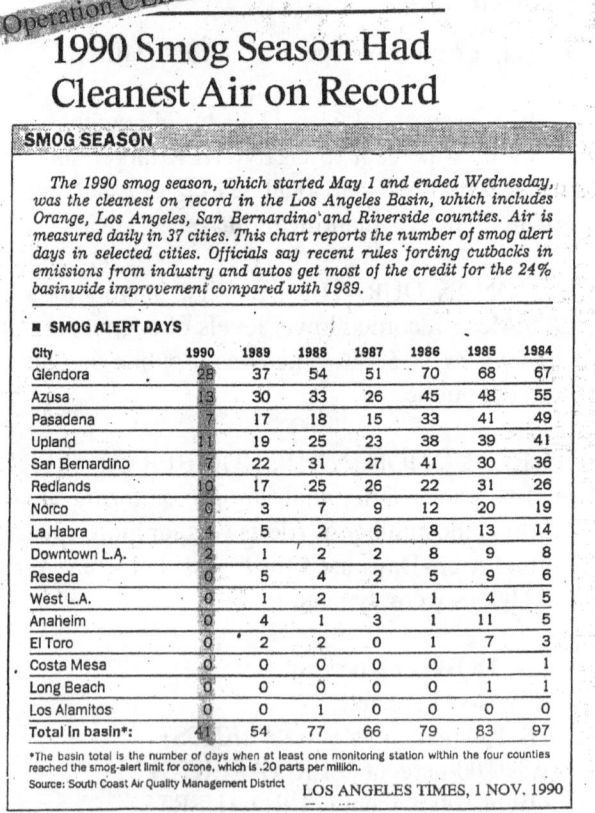

In the same issue of the *L. A. Times*, there's the following quote from Joe Cassmassi (see page 203), which confirms our hard work:

This is mind-boggling... there's really been a dramatic improvement!
—Joe Cassmassi, Senior Meteorologist, South Coast Quality Management District

Overview

The all-around vast reductions in 1990 smog were unprecedented in the smog records. That these reductions were effected from the record low smog levels previously established by Constable's operations in 1989 are convincing in themselves. They are doubly convincing by having been predicted in the Federal CLINCHER filing ahead of the time by the project engineer.

Smog reductions of the 1990 magnitude and scope clearly infer *more power* being used in 1990. The inarguable results jibe with improved equipment, and the use of more than a dozen bases, against a past maximum of four bases. The result of the increased anti-smog power appears objectively as greater smog reductions for longer periods with steeply lessened Alert times and peaks. There were dramatically lower levels of smog overall.

The impression of smog being technically mastered by means of an effective modality intelligently turned against its existence is inescapable. *There is nothing random about what happened during clincher.*

The direct connection in mid-project between General LeMay's intervention, and the subsequent halving of the Riverside smog Alert rate, strongly reinforces this assessment. The huge percentage drop in Alerts, and total suppression of Alerts in districts where Spiders were located, establish a further direct connection to CLINCHER operations.

Those who still wish to argue the existence of the ether, to evade both the essential and the inevitable, should look straight at the facts that glare out at them from the operational record of CLINCHER and the three preceding smog operations that were its basis. Etheric engineering is here – and now. We ignore it to our discredit, detriment and cost.

The AQMD immediately claimed loudly that "tougher regulations" had produced the near-incredible reductions in 1990 smog, as though rules on paper could physically pull all that toxic material out of the air. Trevor Constable had expected the smog bureaucracy to make this kind of assertion, which he classified as preposterous. "The AQMD has never had any compunction about stuffing such rubbish down the public gullet," he said. "CLINCHER will not actually be complete until 31 October 1991, at the close of next year's smog season." He described this as a test of tougher regulations versus etheric engineering.

By the time the 1991 smog season opened in May 1991, Constable expected the AQMD to have many more tougher regulations in force. With the Singapore group completely out of the 1991 scenario, more and tougher regulations should have been able to exceed 1990's records. That never happened, even in unbelievably lucky circumstances. In 1991, thousands of tons of volcanic ash were belched into the world sky by Mount Pinatubo near Manila. This event shaded much of the earth, and gave southern California its coolest-ever summer, a circumstance highly favorable to low smog. Lady Luck plus tougher and more regulations failed southern California in 1991, *when regional smog rose by 12 percent.*

The CLINCHER operation ended for Trevor Constable on a sour personal note. Right after CLINCHER, he transferred a car from Hawaii to California registry. He was charged a cash "smog penalty" of $300 for the transfer. That stung a little. "It's your share of what it costs to fight smog," said the Department of Motor Vehicles bureaucrat. No good deed goes unpunished.

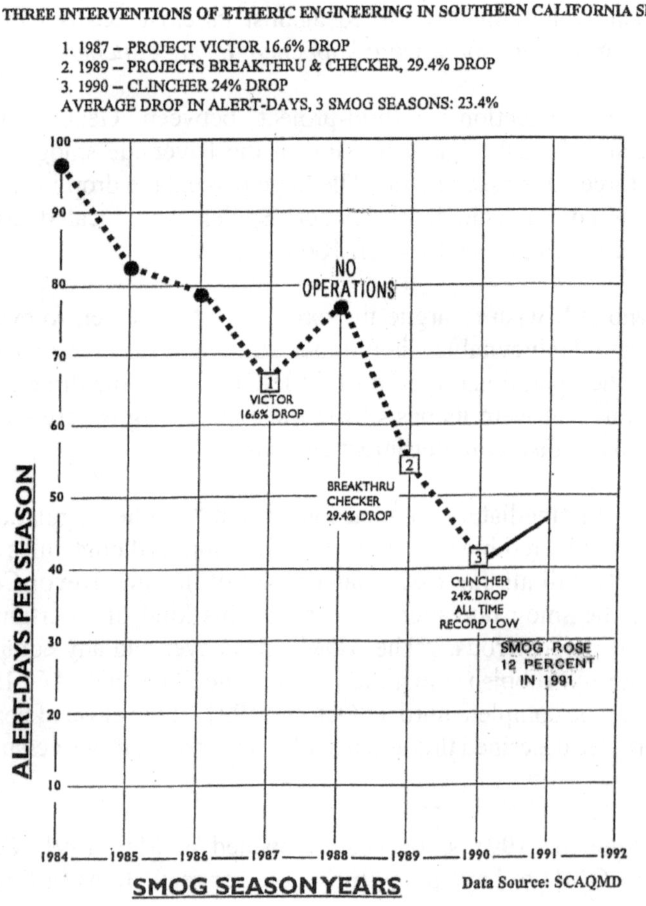

Chapter 21

BAGNOLD'S BLUFF

The Little-Known Figure Behind Britain's Daring Long Range Desert Patrols

Specialist military units of the commando type enjoyed wide vogue during the Second World War, and what little military glamour shone through the conflict was confined almost exclusively to these private armies. They were the stuff of which legends are made. Bold leaders harassing armies with mosquito forces naturally became headline heroes in a war of otherwise inhuman mass effects. Ord Wingate and his Chindits in Burma; Evans Carlson and his Marine Raiders in the Pacific; Mountbatten's commandos; "Phantom Major" David Stirling and his Special Air Services force in North Africa; and on the other side, the unforgettable Otto Skorzeny. The list of famous names is lengthy, and even today they evoke memories of high adventure and piracy. Missing from among them is the brilliant progenitor of all these private armies of modern times, the solder-scientist who conceived and built the first and most successful of them all – Ralph Bagnold.

Ralph Bagnold in a rare photograph taken by his friend Bill Kennedy Shaw during a 1932 exploration trip in Libya's Great Sand Sea. During his private pioneering expeditions in the 1930s he learned the techniques of driving motor vehicles over the immense dunes of northern Africa, and of surviving in the pitiless desert – experience that proved invaluable in organizing desert patrols in 1940. With General Wavell's backing, Bagnold's daring patrols bluffed Italy's Marshall Graziani into halting his drive to the Suez Canal. Wartime security suppressed the story of Bagnold's history-changing achievement.

This tough-minded yet visionary Englishman played a decisive part in bringing the Allies through the serious crisis precipitated by Italy's entry into the war. The loss of the entire Middle East was an imminent possibility. The dramatic, unexpected flanking diversion provided by Bagnold's long range patrols – operating across the mountainous, scorching dunes in the interior of Egypt and Libya – tipped the strategic balance against the Axis. Military units had never penetrated these vast, unmapped wastes before, and First World War patrols had gone no farther than their fringe, where they recoiled for the impassable barrier of the giant dunes. Formal military thinking on North African topography routinely took its cue from this experience. The dunes were deemed to be impassable. The success Bagnold achieved in the teeth of these and other orthodox military conceptions opened many minds in the Allied high command, paving the way for numerous specialist units that followed.

The successful ones were built upon the foundation that Bagnold laid. He established the fundamentals of all small force success – planning, organization, the right equipment and communications, and a human element of exceptional quality. Adherence to these fundamentals could produce results out of all proportion to the size of the force, and with minimum casualties.

Today the ability of a small, highly trained unit to penetrate to the heart of any country on earth has to be taken into account in protecting key leaders in the event of war. The Assassins of the twelfth century may have been the originators of this concept, but it was Ralph Bagnold who first showed in modern times what an elite and resolute small force could achieve in upsetting the strategy of armies. His achievement had its origin in a seemingly useless peacetime hobby that the English adventurer shared with a few friends. How he turned this hobby into a superior instrument of war and was then hidden by the sheer bulk of the commando heroes who came later, is an example of historical caprice hardly rivaled in our time.

A professional soldier who entered the British Army as an engineer officer through the Royal Military Academy at Woolwich, Ralph Bagnold served in the trenches in the First World War. Posted to Egypt in 1925 as a signals officer, he found himself among a group of kindred spirits sharing a combined officers' mess with the Royal Tank Corps. He began experimenting with the cross-country potentialities and endurance of the Model T Ford, taking these rugged early cars over rough ground and sand drifts where no car had previously ventured. While other officers spent their time at Gezira Sporting Club or enjoying the fleshpots of Cairo and Alexandria, Bagnold and his friends used their weekends and periods of local leave to make adventurous journeys in the desert. They probed eastwards to Sinai, Palestine and Jordan before made-up roads existed. Their leader by free acknowledgment, Bagnold's enterprise, ingenuity and intelligence were the driving force behind these pioneering expeditions.

Unshaven in their informal desert garb hundreds of miles from civilization, Bagnold and his friends might have been considered highly unconventional by those who were content with more mundane recreations. They were intelligent, educated men indulging a common passion for the desert. Their numbers regularly included two young officers of the Royal Tank Corps, Guy Prendergast and Rupert Harding-Newman. Both were expert drivers, and Prendergast was also an enthusiastic airman at a time when flying was still a rare skill. Later they were destined to turn their journeys with Bagnold to good military account, although at the time, the far-ranging journeys were merely a hobby.

Growing experience and confidence in his own logistics and specially-designed equipment turned Bagnold's mind inevitably westwards to the frightening immensity of the Libyan Desert – the most arid region on earth. Roughly the size and shape of the whole Indian peninsula, its strange, wind-sculptured wastes, as rainless and dead as the moon, were largely unmapped and untrodden by man or beast since prehistoric times. Scorching, vast and silent, it presented an irresistible challenge. Intrigued by the prospect of conquering this desert of deserts, the English explorer began planning a new adventure.

Could a small, self-financed party of six men, in three of the new Model A Ford cars, penetrate the Libyan Desert as far or perhaps even farther than previous expeditions? The most recent exploration effort had been made by the millionaire Prince Kemal el Din, with a fleet of caterpillar trucks supported by supply trains of camels. Three Model A Fords seemed a puny expedition by comparison, but Bagnold felt that perhaps sheer size and resources were not the key to success. Might not a small party even succeed in crossing the enormous dune field of the Great Sand Sea? The width of that barrier was unknown, but it separated Egypt and Libya for five hundred miles from north to south. Prince Kemal el Din had judged the Great Sand Sea to be utterly impassable.

Despite this first-hand judgment by a contemporary explorer, Bagnold resolved in 1930 to try to conquer the dune barrier. His party included two British officials on leave from the Sudan civil service: Douglas Newbold, permanent head of the government, and Bill Kennedy Shaw, archaeologist and botanist. Both were Arabic scholars and experienced camel travelers, and both were burning with enthusiasm to explore the mysteries of the Libyan Desert, legends of which abounded in ancient Egyptian records and in Arabic literature.

Bagnold's planning and intuitive path-finding succeeded. The bold little group discovered a single practicable route for light cars over range upon range of towering sand dunes. In their Model A Fords they covered some four thousand miles of unknown country before returning to Cairo in triumph. The Sand Sea route was retraced and mapped in detail shortly afterwards by Patrick Clayton, a tough, restless Irishman and expert cartographer employed by the Egyptian Desert Survey. Clayton's grey hairs belied his drive, versatility and skill, qualities which earned Bagnold's respect and friendship.

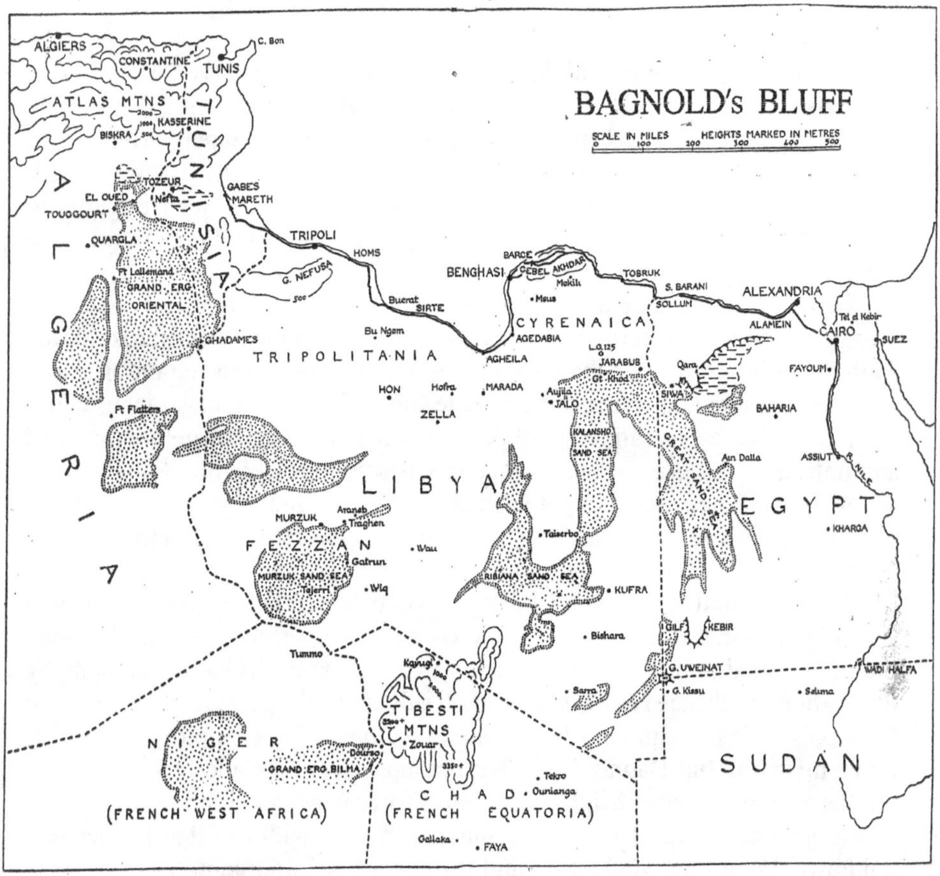

Typical straight-line distances from Cairo traversed by Bagnold's first long range patrols. To Murzuk 1100 mi., to Kufra 780 mi., to Uweinat 800 mi., to Chad Province 1000 mi. *Over-the-ground distances greatly exceeded these mileages.*

After this successful penetration of Inner Libya, the Royal Geographical Society supported additional and still longer journeys. The primary exploration of the region was under way, but Bagnold's interest had meanwhile been seized by the sands in a manner quite different from that of a conventional explorer. Fascinated by the extraordinary symmetry and geometrical regularity of the great dunes, he found that little was known to scientists about the formation and movement of these vast natural barriers. Retiring from the army, Bagnold turned scientist and embarked on laboratory research into sand movement. He wrote a treatise entitled *The Physics of Blown Sand*, which earned him election to the elite Royal Society of London – an almost unique distinction for a service officer with no academic qualifications beyond a Cambridge BA. He occupied himself with his scientific work in communications, hydraulics and fields connected with sand such as beach formation, until the outbreak of war in September 1939.

Major Bagnold was immediately recalled to the army. Ignoring his unique talents and specialized experience, the British Army bundled him aboard a troop ship bound for Kenya – a country of which he knew nothing. The prospect of his years of desert experience going to waste was discouraging, but he could do no more than obey orders.

Fate intervened in the form of a mid-Mediterranean collision involving his troop ship. The vessel was so badly damaged that its passengers were disembarked at Port Said, where they would be required to wait at least a week for another ship. Seizing the chance to visit his many friends in the capital, Bagnold caught the first train to Cairo. A sharp-eyed reporter for the *Egyptian Gazette* spotted the greying major in Shepherd's Hotel, the famous social mecca of British Army officers in those days. The reporter knew all about Bagnold's prewar desert journeys and began putting two and two together. In his column *Day In, Day Out*, he briefly reviewed Bagnold's past achievements for his readers, and ended his column with the following observation:

> Major Bagnold's presence in Egypt at this time seems a reassuring indication that one of the cardinal errors of 1914-18 is not to be repeated. During that war, if a man had made a name for himself as an explorer of Egyptian deserts, he would almost certainly have been sent to Jamaica to report on the possibilities of increasing rum production, or else have been employed digging tunnels under the Messines Ridge. Nowadays, of course, everything is done much better.

Square peg Bagnold was, "of course," on his way to a round hole in Kenya, true to the British Army tradition that the newspaperman had criticized. The course of the North African war was nevertheless to turn on what the reporter had written about Bagnold in the *Egyptian Gazette*. General Sir Archibald Wavell read the thumbnail sketch of Bagnold's desert career in *Day In, Day Out*, and thus learned of the explorer's presence in Egypt.

Although Wavell had no official status in the Middle East at that time, he was working behind the scenes on preparations for the inevitable expansion of the war in that theatre. The so-called "Phony War" was in progress in Europe after Germany's conquest of Poland. The Battle of France still lay in the future. Italy was not yet in the war, and the open appointment of an eminent soldier like Wavell to command the Middle East might have been seized on by Mussolini as a provocation. General Wavell had therefore been sent out from England *sub rosa*, to plan for Italy's entry into the war, or for a German thrust through the Balkans, or for both together. As if to emphasize his unofficial status, Wavell occupied a small office in the attic of the bulky

HQ building of British Troops Egypt (BTE), the peacetime garrison force commanded by General Sir Henry Maitland "Jumbo" Wilson. Bagnold was completely unaware of all these arrangements when he began visiting old army friends.

The major's first call was at the office in the same building of his old friend and contemporary, Colonel Mickey Miller, then chief signal officer of BTE. Miller's face lit up as Bagnold appeared. "Just the man," he said. "Wavell wants to see you." "Wavell?" said Bagnold, "What's he doing here? I thought Jumbo Wilson was in command." Miller put his fingers to his lips in a gesture of silence. "Hush," he said. "Wavell isn't supposed to be here. Jumbo's our boss. Wavell has no authority to interfere. But he knows everything and everybody. He's planning something big and he's collecting people – people who *know* things. You'll certainly be transferred here, Ralph. Come on. I'll take you upstairs."

As they climbed up to the attic, Bagnold's puzzlement grew at the modest quarters assigned to such a senior general. From Mickey Miller came a quick aside as they reached Wavell's office: "He's got a glass eye, you know. So be careful to look at the good one."

The interview was brief. The one very bright eye, set in a wrinkled, weather-beaten face, looked Bagnold over. The general spoke quietly.

"Good morning Bagnold. I know about you. Been posted to Kenya. Know anything about that country?"

"No, sir."

"Be more useful here wouldn't you?"

"Yes, sir."

"Right. That's all for now."

As Bagnold walked out, he pondered on the inscrutability of that remarkable face. Was it grim, or smiling at the prospect of some half-formed plan? Even in those brief moments there was an impression of quiet power about Wavell.

Two days later a cable from London transferred Bagnold to Egypt, and was followed by a local posting to a signal unit of Major General Hobart's Armored Division at Matruh, on Egypt's Mediterranean coast. He was back in the desert again. Cancellation of his Kenyan assignment was like a redemption, but in later years he would marvel at the delicate pinions of maritime collision and newspaperman's acumen on which his destiny had turned.

Within a few weeks, his geographically broader outlook grasped the alarming weaknesses of the defense situation should the huge Italian armies in Libya and Ethiopia attack the Nile states of Egypt and the Sudan. The one British armored division in North Africa, newly formed and crucially short of transport, would be put to its limit to defend the 60 mile-wide "Western

Desert" – the maneuverable coastal strip between the Mediterranean and the northern edge of the great sands. A major Italian thrust to seize the Nile Delta was certain to come from Italian Libya eastwards in the event of hostilities. Five hundred miles to the south, the Italians were known to maintain a garrison at 'Uweinat on the Sudan border, well beyond the southern limit of the Sand Sea. Bagnold knew this country well. From 'Uweinat it was only 500 miles eastward to the Nile over a sand sheet of billiard-table smoothness. A strong mobile column could cover this distance in two easy days, seize the Aswan Dam, isolate the Sudan and hold Egypt to ransom. Bagnold knew that this situation would be readily apparent to at least one man on the Italian side.

The major's mind turned to his Italian counterpart, Colonel Lorenzini, a man of vision, leadership and daring. Bagnold had met Lorenzini in the remote desert eight years previously and had been deeply impressed by his quality. Lorenzini would instantly grasp the situation in the same way as Bagnold, with all its potential for conquest. If the Italian high command had kept Lorenzini in Libya, surely they would be listening to him now. Complicating the situation and heightening its menace was the lack of aircraft for reconnaissance. The British had no machines available of sufficient range to fly south and investigate Italian intentions.

Summarizing the situation on paper, the analytical Bagnold outlined a suitable establishment for such patrols. He added a note suggesting that since no suitable army vehicles existed, it was high time to begin experimenting on a modest scale with half a dozen selected modern commercial vehicles. He made three copies of his proposal, and gave the original to Major General Percy Hobart to read. The hawk-faced "Hobo" was a leading practical pioneer of modern armored warfare, who had risked his career in the cause of strategic mobility, and was in no doubt as to the validity of Bagnold's proposal: "I entirely agree," said Hobo, "and I'll send this on to Cairo. But I know what will happen. They will turn it down."

Hobart was right. General Wavell had not yet come out of his attic. The Cairo brass lived in the peacetime routine of an internal security force stationed in Egypt since 1870 – an atmosphere lethal to any innovations such as what Bagnold was now proposing. The formal Cairo view was that the desperate lack of defense troops and equipment made it essential not to provoke Italy in any way. Mussolini had a quarter of a million troops in Libya and a quarter of a million more in the south. He was still sitting on the fence. Roving patrols like Bagnold's – even if they were feasible – might tip Mussolini into war. But this was only the formal view.

The real reason for the rejection of Bagnold's proposal lay in the ignorance of the Cairo brass about the desert on whose edge their own HQ was located. Fear was the inevitable concomitant of this ignorance. One senior staff officer warned Bagnold that if he took troops into the desert where there were no roads "you'll get lost." On the officer's wall hung a map of Egypt's western frontier that was dated 1916. Detail in this faded out with the words, "limit of sand dunes unknown." Comments on Bagnold's suggestion of taking patrols across the 150-mile wide Sand Sea ranged from "ridiculous" to "madness."

Physically a wiry man, without an ounce of spare flesh on him, Bagnold had the moral and mental fibre to match his physical resilience. He decided to try again. He showed the second copy of his patrol force proposal to General Hobart's successor, after "Hobo" had been kicked out of the army to become a Home Guard corporal. The new armored division commander also approved of the plan and recommended it to Cairo. Again it was rejected. There were mutterings among offended brass-hats about this second attempt, and the "bloody nerve" of that major out at Matruh.

Shortly afterwards Bagnold went to Turkey in civilian clothes as the signals member of a small reconnaissance mission, sent at the invitation of that nervous and neutral government. When he returned to Cairo, he found the scene transformed. Wavell had come out of his attic. He was now Commander-in-Chief Middle East, a military overlord with responsibility stretching from the Burmese border to West Africa, and from the Balkans to South Africa. A new headquarters, GHQ Middle East, was being set up in a different and cleaner part of Cairo, and an all-new staff consisting largely of officers fresh out from England was being assembled. The atmosphere was freshly alive.

Bagnold was appointed an aide to General Barker, the new Signal-Officer-in-Chief. Involved in the urgent improvisation of communications for Wavell's gigantic and complex command, Bagnold forgot the desert until June 1940 brought crisis. France collapsed. Italy declared war. Both the Mediterranean and the Gulf of Suez were closed to shipping, virtually isolating the Middle East from a Britain itself set upon by a fleet of U-boats and the Luftwaffe. The threat Bagnold had foreseen with Italian entry into the war was now a stark reality.

Marshal Graziani's 15 divisions – a quarter of a million fighting men – would soon start rolling eastwards along the coast road towards Egypt and the Suez Canal. The Duke of Aosta's similarly massive army in Ethiopia posed a similar threat, pincering in on the Sudan and Egypt from the south. Wavell's immediately available defense forces were outnumbered ten to one. Reinforcements were coming, but with the Mediterranean closed, their arrival and deployment might be delayed for months. There were no war reserves of weapons or equipment. The situation seemed desperate.

The hour was late and Bagnold acted. He dug out the last copy of his earlier patrol force proposal and persuaded the head of the Operations Staff to place it personally on the commander-in-chief's own table. Reaction was immediate. Within an hour, Bagnold was again alone with Wavell.

This time there was no oppressive attic office, lack of authority, or doubt about the crisis that was being confronted. The great man on whom so much now depended sat calm and relaxed in his chair, the one eye bright as before. His greeting set Bagnold at ease, for Wavell acted like a shy man welcoming a friend for a quiet chat. He indicated the rumpled paper lying on his desk. "Tell me about this, Bagnold. How would you get into Libya?" Bagnold walked over to a modern map of Western Egypt hanging on the wall, and his finger stabbed and then moved laterally. "Straight through the middle of the Sand Sea, sir. It's the most unlikely place. The passage is here, due west of Ain Dalla. I've been along it, and I'm sure it will go all right, sir. And the going is good on the other side, what Clayton saw of it." "

The C.-in-C.'s weather-beaten face was impassive. "What would you do on the other side?" he asked. "We would go far enough west to cross both the southerly routes to Kufra Oasis and 'Uweinat. By reading the tracts, we could tell what recent traffic had been along them – the direction of travel and type of vehicle."

Wavell's expression remained unchanged. "What are the risks?" "Two, sir. First, the weather. No Europeans have been into the sands in summer. If a south wind gets up, it'll be pretty hot. How hot no one knows. Second, this map of yours, sir. You see the passage across the Sand Sea is printed on it, and it's been on sale in Cairo for years."

Wavell gave a comprehending nod. "You mean they might be waiting for you at Clayton's cairn on the other side?" "Yes, sir. But it's a bleak place for Italians to live at – no water, no life, no shelter and far from anywhere. It's a reasonable risk to assume they won't be there." "What about *your* wheel tracks, Bagnold? They last for years." "Over gravel country, yes, sir, but it's very difficult to follow wheel tracks from the air. The aircraft goes too fast. If you fly low enough to see the tracks, they suddenly jink sideways under the fuselage and are lost. Our tracks over the dunes, of course, would disappear with the first bit of wind."

The C.-in-C. leaned forward a little in his chair, still inscrutable, but obviously interested. "And if you find there has been no activity along the southerly routes, what then Bagnold?"

"How about some piracy on the high desert?"

Wavell's face changed sharply. For an instant Bagnold feared he had gone too far. He had been too flippant with the C.-in-C. But the wrinkled face had creased now into a broad grin, the eye was very bright indeed and his whole head could have belonged to a pirate captain.

"Can you be ready in six weeks?"

"Yes, sir."

"Any questions?"

"Volunteers and equipment, sir."

"Volunteers are a job for British Troops Egypt. I'll see that General Wilson gives you every help. Equipment? Hmmm, yes. You'll meet opposition."

Wavell reached out and pressed a button. Expecting a clerk or orderly to enter, Bagnold was astonished when the bell was answered immediately by a lieutenant-general. He was Sir Arthur Smith, Wavell's chief of staff. "Arthur," said Wavell. "Bagnold seeks a talisman. Get this typed out for me to sign, now." The C.-in-C. then dictated the most amazing order that Bagnold had heard in his military career: "Most Secret. To all Heads of Branches and Directories. I wish any demand made personally by Major Bagnold to be met urgently and without question."

Wavell turned now to Bagnold. "Not a word of this must get out. There are some sixty thousand enemy subjects of all classes loose in Egypt. Get a good cover story from my DMI [Director of Military Intelligence]. When you're ready to start, write out your own operation orders and bring them direct to me."

This was absolute *carte blanche* – regardless of desperate equipment shortages.

Leaving the C.-in-C.'s office still hardly believing his ears, Bagnold pondered the sudden reaction and quick decision at the suggestion of piracy. Why had that word precipitated action? He reviewed what he knew of Wavell in search of an answer. A brilliant, mobility-minded strategist, Wavell was a student of foreign armies and the mentality of their leaders. He was also a poet and author. A member of Allenby's staff in the masterly Palestine campaign of the First World War, he was even now finishing a biography of the former chief. There was something else about Wavell – his grasp of strategic deception. He had made it a science. That must be it. The old man was planning an immense bluff to play for time!

The next six weeks were the most demanding and challenging of Bagnold's life. A new and untried type of armed force had to be created from nothing, trained for operations never previously attempted and introduced to a hard and novel way of life – all in a few short weeks. Success would depend on combining Wavell's talisman with a clear-cut plan and a knowledge of which button to push in the giant HQ machine. Bagnold threw all his energy into the task.

He would need the help of his prewar companions. Rupert Harding-Newman was the only one locally available in Cairo, serving as a liaison officer with the non-belligerent Egyptian Army. Guy Prendergast could not be brought from Britain. The archaeologist Bill Kennedy Shaw was curator of the Jerusalem Museum. Pat Clayton was on a surveying job in the wilds of Tanganyika. Shaw's release by the Palestine government was arranged, and Clayton was located by the Tanganyika government and bundled aboard a special aircraft for Cairo. Shaw and Clayton were both in Cairo within three days of Bagnold's request for their services, and both were put into uniform and commissioned as army captains immediately.

Bagnold and Harding-Newman meanwhile went shopping round the Cairo truck dealers. After trying out several types and makes, they settled on a one-and-a-half ton commercial Chevrolet with two-wheel drive. All the dealers in Egypt could supply only 14 of these vehicles. Hearting-Newman persuaded the Egyptian Army to part with another 19. The Ordinance Directorate workshops put aside all other jobs to modify these commercial trucks to the specifications rapidly drawn up by Bagnold and Harding-Newman. Cabs were cut off, windshields were discarded, chassis and springs strengthened and gun mountings installed. Open bodies were built to Bagnold's design, into which modular ration and fuel cases fitted precisely for perfect stowage and prevention of fuel loss.

There were to be three patrols, each with two officers and 28 men, carried in ten unarmored trucks, with a lighter vehicle for the patrol commander. Each patrol would carry a two-pounder gun, and each truck anti-tank rifles and machine guns mounted for all-round fire. Crews were to be individually armed with pistols and rifles. Bagnold specified endurance requirements unprecedented in military history. His tiny force was required to navigate and operate entirely on its own, far beyond the reach of any help or supplies. Each patrol had to be self-contained for fuel, water, food, ammunition, spare parts and radio for 1,500 cross-country miles – the equivalent of 2,800 road miles – and for at least two weeks. Range could be increased by making double journeys to form supply dumps.

One of Bagnold's early Long Range Patrols is inspected in Cairo in 1940 before departure on operations across the desert and dunes of Eastern Libya. At first made up largely of New Zealand volunteers, the patrols' main tasks were to reconnoiter far west of the main fighting lines to report on enemy movements and dispositions. Later known as the Long Range Desert Group (LRDG), and greatly expanded, the patrols were active throughout the North African war. The LRDG was organized in a dozen truck-borne patrols, with ten trucks to a patrol and some six men per truck. ITs tactics and administration were fluid. More than 50 of its members were decorated for gallantry, and only 16 were killed. "Considering its size," concluded The Historical Encyclopedia of World War II (1989), *it [the LRDG] exercised a wholly disproportionate influence on the desert war."*

Every pound of weight mattered. Only absolute essentials could be carried. For this reason Bagnold had chosen the simpler two-wheel drive trucks. Gearing for the four-wheel drive vehicles added considerable weight and fuel consumption was higher. All the innovations introduced and proved by Bagnold for his expeditions years before were now brought into full use. Every vehicle was equipped with a pair of portable steel channels for "unsticking" from soft sand. These channels were placed under the rear wheels of a truck stuck in soft sand, and provided a sure method of extricating the heaviest vehicle.

Patrol trucks of Bagnold's LRDG like this one could cross desert dunes impassable to conventional units. This enabled them to carry out comprehensive surveillance of Axis supply lines hundreds of miles behind the front, and to harass the enemy's rear and communications. Their intelligence-gathering activities and raiding operations immeasurably aided the Allied cause.

Bagnold had also discovered how to conserve water. The regular radiator overflow was blocked up, and another hole made in the top of the radiator. From this hole, a hose was led to a vertical pipe reaching to the bottom of a two-gallon can bolted to the front fender. With the can initially half full of cold water, steam and boiling engine water which would otherwise escape were collected in the can. Half a minute after the engine was stopped, condensation caused all this water to be sucked automatically back into the radiator. As a precaution against long-continued boiling, possible when plowing through soft sand in a hot following wind, scalding water would ultimately spurt up from a vent in the can top on to the driver, forcing him to stop. With this ingenious Bagnold invention, trucks would need no extra water in hundreds of miles of hard desert driving. (By the end of the 20th century, virtually all passenger cars came fitted with versions of the device Bagnold invented in the 1920s.)

Navigation had long presented a problem in trackless, uncharted desert, mainly because a magnetic compass was unreliable in a moving car. In 1928 Bagnold had produced a simple, homemade sun compass by which the true bearing could be read accurately and quickly even when the vehicle was bumping over rough ground. The navigator could thus make a continuous record of bearing and mileage regardless of the course changes made by his leader to avoid obstacles. This record permitted a reliable plot to be made of the preceding route at every halt. The British Army had not adopted this type of sun compass, invented by one of its own engineer officers. The Egyptian Army had nevertheless seen the merits of Bagnold's compass, and had fortuitously received a recent consignment from the London makers. Bagnold was able to borrow some of his own inventions from the Egyptians.

Each patrol needed a theodolite, a surveying instrument used for accurate position-fixing by the stars at the end of each day's run. The British Army could produce but one in the entire Middle East. A second was borrowed from Bagnold's old friend George Murray, director of the Egyptian Desert Survey and Pat Clayton's former chief. A third theodolite was located at Nairobi in Kenya and flown to Cairo.

Murray also helped by printing map sheets of Inner Libya on the same useful half-million scale as his own Egyptian maps. Although virtually blank, these sheets were essential for plotting courses and recording information along the route. The British Army had not yet devised any rational method for carrying the great number of maps it needed, toting them about in clumsy bundles of springy rolls, which filled a three-ton truck. Bagnold's prewar expeditions had conquered this problem, since maps had to be stowable in the confined spaces of a Model A or Model T car. Bagnold had devised light plywood portfolios in which maps were packed flat. A supply of these portfolios solved the same problem now.

The health needs of the personnel demanded special consideration. Men would be exposed for long periods to desiccating heat by day while on a limited water ration, and to a large temperature drop after sunset in the desert. Prewar experience had demonstrated the adverse physical and psychological effects of a monotonous diet of army preserved rations under such conditions. Bagnold and his committee of old hands decided that a special ration scale was essential. Pat Clayton unearthed a nine-year-old copy of the *Geographical Journal* in which Bagnold had recorded a diet found suitable by his prewar expeditions. To the Supply Directorate, departure from the London-ordained ration scale was sacrilegious. There was resistance. With Wavell's talisman and the full backing of the Director of Medical Services, Bagnold nevertheless got his way. The new patrol unit was allowed its special ration scale, including the daily issue of rum, abolished for the rest of the army since the First World War.

Staid British quartermasters raised their eyebrows time and again at Bagnold's unorthodox demands. As the patrols would navigate by the stars over trackless desert, Air Almanacs were requisitioned. Army boots, which would fill with sand, were clearly unsuitable, so Bagnold had *chaplis* – sandals as worn by the Indian frontier tribesmen – made to special order. For protection against wind, sun and sandblast, he chose the flowing traditional head-cloth of the Bedouin. Secrecy precluded buying this headwear in the Cairo stores, so a supply was arranged from the Palestine Police.

Each patrol needed a long-distance radio. Bagnold chose a lightweight No. 11 army field radio he knew to be reliable. Transmitting range of this set was officially 70 miles, but for a short period in each 24 hours it would have a much greater "skip" range predictable according to a schedule varying with season and latitude. The No. 11 thus was less than ideal, but nothing else was available. When Bagnold's patrols were equipped, the last No. 11 radio set in the Middle East war reserve went to his third patrol. When he drew his machine guns, three more remained as the reserve for the entire Middle East. Clearly Wavell was dependent on the success of this bluff.

With his unique knowledge and enormous personal drive, Bagnold conquered each problem as it arose. His friend Bill Kennedy Shaw says of this period: "Bagnold's secret weapon was that he knew the desert and he knew the army – and all the quirks of both." The son of a colonel, his second home was the army and the desert his first love. This proved a winning combination, especially when the time came to turn from equipment to personnel. The imaginative major with unorthodox ideas knew enough about the army not to seek volunteers from among the regular troops. He was a realist. There was no time to unlearn such men of their routine ways. Resourceful, responsible men were needed, with the initiative that formal soldiering all too often extinguishes. His patrol personnel would have to absorb in weeks a mass of desert lore that Bagnold had acquired over two decades. They had to be fighting men, and yet skilled tradesmen, fitters, navigators and radio operators – as well as truck drivers and gunners. Keeping their small self-contained force operating for long periods in remote enemy territory would make heavy demands on their vital powers. They should be men accustomed to the outdoors.

General Sir Henry Maitland "Jumbo" Wilson, GOC [General Officer Commanding] of British Troops Egypt, suggested to Bagnold that he would find the men he wanted in the New Zealand Division. "The commander-in-chief has told me about this job of yours," said Wilson. "Sheep farmers should suit you, I think. I'll sound out General Freyberg. These people aren't very keen on serving with 'pommies,' as they call us, but his division has arrived without its weapons, which were sunk at sea." Wilson set up a meeting.

Armed with a detailed list of requirements, Bagnold went to the New Zealand camp near Cairo. The bulky, battle-scarred "Tiny" Freyberg, with his unsurpassed fighting record in the First World War, was an almost-legendary hero to his own men, and he guarded their fortunes in turn with vigilance. His initial reaction was hostile. He was reluctant to lose to his battalion commanders their best men, for this was in effect what Bagnold was asking. Fate, however, had made him the friend and confidant of Percy Hobart during the latter's struggle for strategic mobility. Hobart's ideas had rubbed off on Freyberg, who was also mobility-minded, and Bagnold's proposal was for mobility on a previously unimagined scale. Freyberg gave in. "All right," he said, "You can have them, but only temporarily mind you." As Bagnold left, Freyberg shot after him, "I shall expect them back."

Freyberg's circular to his division calling for "volunteers for an undisclosed but dangerous mission" produced more than a thousand applicants. Freyberg selected two officers from these, Captain Bruce Ballantine and Lieutenant Steele, and told them to pick their own team. When the little army arrived in Cairo, the modified trucks were just beginning to emerge from the workshops. The New Zealanders' initial suspicion of English officers increased when they were received by a major and two captains who seemed to them to be somewhat elderly, greying gentlemen. Qualms were quickly supplanted by enthusiasm as they learned what they were to do, saw the equipment they were to do it with, and how everything had been thought out in meticulous detail.

Under Bill Kennedy Shaw's instruction, the six navigators-to-be quickly learned how to use the sun compass on the move, to plot dead-reckoning courses and to fix their nightly position by the stars. Unexpected help came from one of the volunteers, Private Dick Croucher, who admitted to being an ex-Merchant Navy Officer with a first mate's ticket. Like many other New Zealand soldiers, he had concealed his qualifications for fear of having to spend the war on a busman's holiday.

Within the six weeks' time limit set by Wavell, Major Bagnold was ready with three patrols. Dumps of supplies had been made at Ain Dalla near the Sand Sea crossing as part of their cross-country driving instruction. Another dump had been made at Siwa Oasis to the north of the Sand Sea, whence Pat Clayton had already reconnoitered a second route into inner Libya. When two trucks and five picked New Zealanders, he had penetrated southwards over the hundred-mile-wide northwestern arm of the sands. Clayton also discovered and crossed another vast dune field, little realizing that twenty years later a rich oil field would be located beneath this barrier.

Wavell came personally to say goodbye to the patrols. The great general obviously loved adventurous enterprises, and his weathered face wore a subtle grin as he looked over his "mosquito columns" as he called them. "The old man looks as if he's dying to come with us himself," said a New Zealand trooper.

On September 5, 1940, the patrols slipped out of Cairo in secret. Lest the delicate sand structure of the passes over the dunes might not stand the disturbance of so many wheel tracks, two patrols commanded by Clayton and Steele drove to Siwa Oasis. They made a double journey south over Clayton's Cairn, the marker built by the surveyor ten years previously. The third patrol, commanded by Captain Mifford, with Bagnold, Intelligence Officer Bill Shaw and Adjutant Ballantine drove to Ain Dalla, to cross the Sand Sea directly from east to west. Graziani's huge Italian army was already advancing along the coast road to invade Egypt. Siwa Oasis and its dump would probably fall into enemy hands shortly, closing the newer route to Inner Libya from the north. Everything therefore depended for future operations on the practicability of the Ain Dalla route, crossing the dune ranges at right angles. The burden of finally proving his conception thus rested squarely on Bagnold's shoulders. If he failed, he would fail Wavell at a time when a bluff might be as effective as fifty thousand soldiers in the field – forces that Wavell did not now have.

The patrol quickly reached Ain Dalla, a deserted artisan spring on the eastern side of the sands, two hundred jolting miles from Cairo. The dump was intact. The absence of fresh camel tracks indicated that its existence was still a secret.

After a short rest, refueling and loading up from the dump, the patrol set out for the great dunes a few miles distant. Bagnold took the wheel in the leading truck. Months of planning and effort focused on this tough, austere Englishman whose insight, energies and vision had brought them all to this moment of challenge. He showed no emotion, but only the same pervasive practical confidence they all had come to respect. Tension mounted as the New Zealanders saw ahead a high rampart of sand stretching away unbroken to both horizons. Driving closer, they found the barrier rising ever higher above the flat and rocky plain over which they were lurching. The dazzling yellow glare struck down from the rampart with ever-increasing intensity, blanching the eyes and so magnifying the steepness of the dune that it seemed like a vertical wall hundreds of feet high. Lumbering towards this frightening mountain in trucks carrying over two tons each, the challenge assumed a terrifying immediacy.

Halting the party some distance away, Bagnold let in the clutch and went rolling forward in his single truck to demonstrate the art of assaulting the dunes. Jamming his foot down to the floorboards he sent the heavy vehicle charging at the barrier with engine roaring and speedometer climbing. At 50 mph the wall of glaring yellow came rushing at them alarmingly. Clutching wildly at the dashboard, the man beside Bagnold let out a terrified bellow. "Christ," he yelled, "you're not going to..." Crashing head-on at full tilt into the dune, all wheel motion suddenly seemed to cease. Nose tipping upwards, the truck rose bodily with its cargo as though hoisted from above. "It's the first hundred feet that matters," barked Bagnold above the engine noise. "If you can rush that without wavering, you're all right – if you know where to go."

With a deft, shockless change of gear the truck continued to glide up a long but gentler slope to the crest, where Bagnold slowed and turned quickly. The men with him saw why. A 50-foot precipice of avalanching sand fell away beside them. Braking very gently, Bagnold brought the truck to rest. He'd done it. A ragged cheer went up from the trucks far below them on the plain.

Seconds later Bill Shaw drove his truck up with the same apparent ease. Alarm turned to confidence as the men saw that once again these veteran Englishmen knew what they were doing. Bagnold waved for the rest of them to tackle the climb. He had lectured them on the art of climbing dunes, but it could be learned only by doing it. Nerves of steel and a cast-iron gut were needed to charge that wall at full throttle. The drivers made the almost inevitable mistake: lack of initial momentum, getting into the tracks of the truck ahead, wild gear-changing at the wrong moment, and trying to restart on the steep up-grade. A chaotic jumble of trucks resulted, their wheels stuck deep in bottomless sand.

Hours of backbreaking labor under the searing sun were needed to free the trucks, reassemble them and get them to the top. Bagnold was everywhere amid the sweating, cursing men, demonstrating, instructing, guiding – a leader by example. One by one the vehicles were extricated, in a grueling ordeal that lasted all day. The sun was setting as the last truck reached the top.

They looked westwards over a new and fantastic world of endless repetitive curves. Solid ground could be seen nowhere, submerged as it was to an unknown depth. Range upon parallel range of giant dunes reached away limitlessly to the far horizon, two or three miles apart, like enormous ocean waves about to break but with their motion frozen. In black shadow the setting sun picked out the deep gulfs between each range of dunes. The Great Sand Sea. No more appropriate name was ever given to any natural feature.

Even Bagnold's close friends of prewar years had felt and respected his deep natural reserve, feeling that nine-tenths of the man was essentially hidden. He withdrew by himself now, and sitting propped against a truck he added up the score in the greatest game he would ever play. Distance covered from Dalla: ten miles; distance to the other side: 140 miles of bottomless sand and 40 more of these monumental dune ranges. The trucks could do it, overloaded as they were. He and Bill Shaw had already proved that. The New Zealanders would soon learn. They were adaptable and remarkably intelligent. He comforted himself by remembering the mess his little party had made of their first attempt here ten years ago – and how they had suddenly got the hang of the new driving art.

Bagnold's main anxiety was the weather. Right now it was good. Temperatures were little more than 100 degrees F in the shade, although things in the sun were too hot to touch. No sand moved in the gentle breeze, but if a "qibli" were to hit them in the Sand Sea, the results could be disastrous. He had once left a truck abandoned for two days, and found it lying on its side at the bottom of a ten-foot-deep scour hole. In a "qibli" the whole surface, now so still, would flow like water. They'd have to keep moving, yet it would be impossible to see through the mist of stinging sand grain, impossible to pick a firm route between dry quick-sands, or to keep men and trucks in sight. The Sand Sea could swallow the patrol under such conditions. Only Bill Shaw among them knew of the weather risk, and he would keep it to himself.

Bagnold began the next day with some short practical lessons. "Once you understand one of these dune ranges," he said, "you understand them all. They're all one family. Look at our tracks... barely half an inch deep. The sand here will take great weight, yet its quite loose. You can run your fingers through it. The reason it's firm is that its surface is streamline to the wind. And the wind has fitted the grains one by one to make the densest possible packing." He pointed to a nearby area. "See that different ripple pattern and the slight change of color? That sand was deposited at random, maybe centuries ago, by avalanching down a dune that has now marched on. Walk over and try it."

Two soldiers walked to the area and sank in nearly to their knees. Once more, the grey-haired Englishman spoke the truth. "Many patches like that," he said, "are covered over like thin ice over water. So keep your speed up, in the hope that your momentum will carry you through. But keep a sharp lookout for sudden precipices like this one. If you get stuck in soft sand, never, never try to get out without putting the channels under the back wheels. Remember that even this firm sand here is loose. Treat it like ice and be very gentle on both brake and clutch."

This learn-as-you-go instruction paid immediate dividends. The patrol covered 40 miles of the Sand Sea the second day. Expertise quickly developed in unsticking bogged trucks, and in foreseeing where the surface was likely to be soft. Confidence increased with skill. On and on they ploughed, until by noon on the third day they emerged triumphant on to a vast, featureless plain of hard, sand-strewn gravel. They were across!

Looking back along their tracks, the men marveled how anyone could find those narrow ascents on firm sand between so many avalanching sand cliffs.

How could any man weave his way so unerringly across that bewildering landscape without a single landmark? How in hell had the major done it? Bagnold overheard some of their incredulous comments and wondered himself how he had been able to pull it off after a lapse of ten years.

Here on the Libyan side of the sands stood Clayton's Cairn, the surveyor's professionally built pillar of loose stones. No man or animal had since visited the God-forsaken spot where they now stood. The only tracks were Clayton's of ten years previously – still visible in the gravel. Bagnold's underground route into Libya was still a secret from the enemy. While waiting for Clayton's two patrols from the north to rendezvous, Bagnold mounted a return journey to Dalla for supplies. The men's newly acquired skill showed in the scant seven hours they needed for the trip each way. When Clayton arrived, they had a substantial supply dump at Clayton's Cairn, and the complete mosquito army stood ready for action. Bagnold's bold concept had been vindicated in its most critical phase.

A military force could cross the Great Sand Sea, and in this brand-new fact lay considerable strategic possibilities. The inner desert no longer provided a defensive flank to an enemy attacking along the coast, but instead lay open before Bagnold's little force. The slender north-south lines of communication from the Mediterranean coast to Graziani's bases at Kufra and 'Uweinat in the far south could be harassed at will.

On September 13, 1940, Graziani's Libyan Army crossed the Egyptian frontier on its eastward advance towards Cairo and the Suez Canal. On that same day, Bagnold launched a two-pronged probe westwards into the heart of Libya. Mitford's patrol struck westwards across five degrees of longitude, burning the stocks of petrol found on the chain of landing grounds along the Kufra air route. They examined the motor tracks leading south, and kidnapped a small motor convoy complete with vehicles, supplies and official letters.

Pat Clayton meanwhile struck south-westwards, passing between Kufra and 'Uweinat mountain, right across southeast Libya to make contact with an astonished French outpost of Chad Province in French Equatorial Africa. Skirting the enemy garrison at 'Uweinat, the patrols rendezvoused in the desert and returned via Ain Dalla to Cairo. The prisoners and captured letters were handed over to Intelligence, and proved to be a mine of information for General Wavell. From Clayton's Cairn to Dalla, the patrols had travelled 1,300 miles completely self-contained. 150,000 truck miles had been covered without a single serious breakdown. This was only the beginning.

Other more aggressive raids quickly followed. Enemy desert outposts in the north were bombarded and destroyed, their garrisons routed or taken prisoner. Simultaneously the garrison at 'Uweinat was attacked 500 miles to the south. A collection of aircraft was destroyed on the ground, and a large dump of bombs and ammunition blown up. The attackers seemed to emerge from the fourth dimension to strike and vanish like lethal ghosts. They appeared, struck and disappeared at widely separated points, seemingly within hours of each other. British radio monitors in Cairo and elsewhere intercepted enemy messages of alarm and cries for help pouring into Graziani's headquarters from all over eastern Libya.

All Graziani's plans for the conquest of Egypt were based on the assumption, backed by his intelligence reports, that he faced only weak forces. Quick victory and occupation of the Nile Delta were anticipated within a few weeks. Yet within a few days of his first battalions crossing the Egyptian frontier, he began getting these disturbing reports of attack – from a direction he believed to be completely secure. The British seemed to be everywhere, operating at incredible distances from their base. These assaults gave the war situation a new dimension. These far-ranging forces might attack his vital rearward lines of communication. Graziani ceased believing his intelligence reports and their central theme of British weakness. In overwhelming strength, the massive Italian army halted its advance. Wavell's bluff was beginning to succeed.

Exploiting the situation to the full, Wavell ordered the number of patrols to be doubled. Twice as much piracy would spring from the additional patrols. From Long Range Patrols, the force was given a new designation, Long Range Desert Group (LRDG). The new distinguishing badge, showing a scarab riding a wheel, made its wearers among the most respected soldiers in the Middle East. Volunteers from the Brigade of Guards, Yeomanry regiments, the Rhodesian Army and the Indian Army joined the pioneer New Zealanders.

With the doubling of patrols came a second carte blanche from Wavell: Stir up trouble anywhere in Libya where the enemy can be harassed, attacked or shaken. Bagnold promptly obliged. He mounted an attack on Murzuk and its landing strip 1,100 miles as the crow flies from Cairo and 1,400 miles over the ground. Murzuk and back was far beyond the maximum range even of Bagnold's patrols, but supply dumping by the Free French in Chad under the command of Lieutenant-Colonel d'Ornano provided the necessary extension of range. (Colonel d'Ornano's "price" for supply assistance was to be permitted to participate in the Murzuk raid. He was killed in action there.)

The brilliant Free French military commander, Leclerc, stimulated by what he had heard of the capabilities of the British patrols, soon afterward resolved to capture the Italian stronghold at Kufra Oasis. Kufra was too tough a nut for Bagnold's small force to tackle alone. By a miracle of improvisation, Leclerc overhauled and equipped sufficient local transport to carry a battalion of native Chad troops and two 75-millimeter field guns, together with supplies for the double journey of a thousand miles. The attack on Kufra, backed by Bagnold's patrols, finally cleared the enemy from the whole interior of eastern Libya. The Murzuk and Kufra strikes were timed to coincide with Wavell's counter offensive in the Western Desert. With the time that Bagnold's force had won, Wavell had built up his strength, and by February 5, 1941, he had smashed the Italian Army.

From this time until the end of the North African war, at least one patrol of the Long Range Desert Group was always behind enemy lines. The unit doubled in size yet again. An LRDG "private air force" was added, in the form of two WACO monoplanes purchased from an Egyptian pasha, which aided communication with headquarters and evacuation of the wounded. The LRDG guided and carried commando units far behind the front to carry out daring raids. With its unrivalled travelling and navigating abilities, the LRDG could place espionage agents at the very gates of Axis-held strong-points almost anywhere in North Africa.

LRDG patrols themselves razed airfields in daring nocturnal raids, destroying hundreds of aircraft on the ground between 1940 and 1943. Beating up Axis supply convoys and mining roads hundreds of miles behind the front was their steady war routine. The LRDG set up "road watch" patrols, often lying within ear shot of the enemy and reporting every vehicle, weapon and tank that passed by. This precise intelligence of Rommel's supply position was one of Montgomery's vital tools in the ultimate defeat of the Desert Fox. When Ritter von Thoma, Rommel's deputy, was captured in the Battle of Alam Halfa just before El Alamein, the German general was shocked to learn that Monty knew more about the supply status of the Afrika Korps than he did. Most of this information reached Monty via LRDG road watch patrols.

Ritter von Thoma (left), Rommel's deputy, leaves the tent of Field Marshal Montgomery (right) after his capture in November 1942.

The patrols continued to penetrate Axis territory about as much as they chose. In the immensity of the desert, their vehicles were rarely spotted. Bagnold's original concept, his detailed development of it, and his far-seeing organization had transformed the inner desert from a textbook "defensive flank" into a serious liability to the enemy.

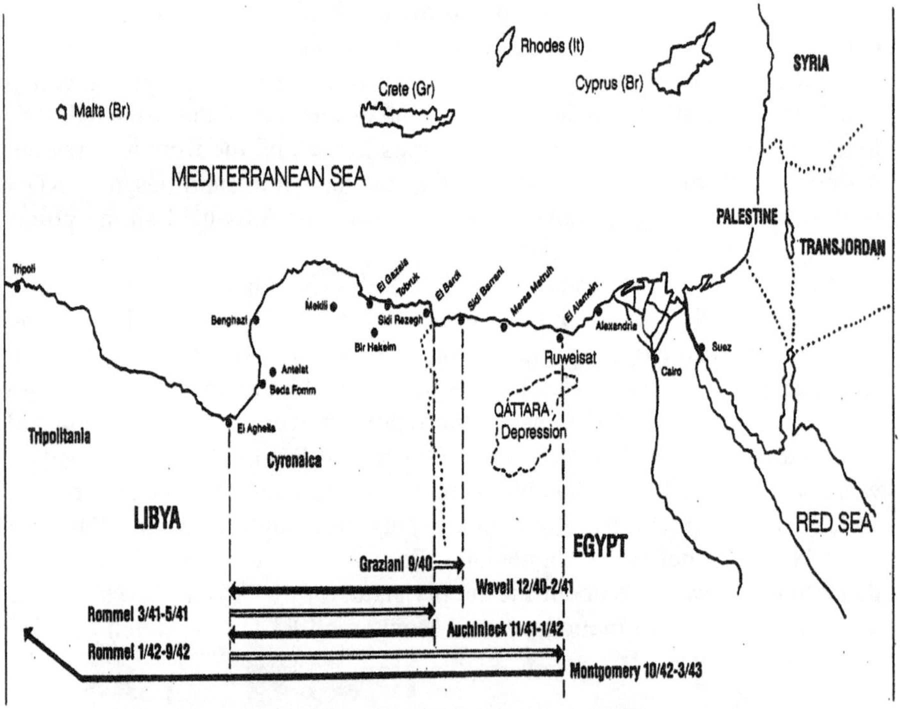

Back and forth across North Africa, 1940-1943.

In action against the Axis forces in North Africa from first to last, the LRDG proved to be the most original, boldly conceived and brilliantly organized "private army" of the war. The success of Bagnold's patrols helped break down official opposition to those commando-time formations, specialist units and "private armies" that fulfill novel and essential roles for which orthodox forces are neither trained nor equipped. The commando idea had been current for half a century or more, but its modern potentialities under special conditions had never been seriously considered. Those at the top seldom possess the special knowledge and experience to judge the probability of success. Luckily for the Allies, and perhaps for the world, Wavell was a commander willing to take risks. Without the stunning success Bagnold achieved, it is doubtful if some of the later private armies would have been authorized.

Unfortunately for Ralph Bagnold, the modernizer of this kind of auxiliary warfare, the *modus operandi* of his unique force had to be concealed in wartime from the enemy. Security blocked all details of its size and capabilities. Writing about the LRDG was initially forbidden and later heavily censored. For this reason, the LRDG was far less well known in wartime than other auxiliary forces such as Carlson's Raiders, Wingate's Chindits, Stirling's Parashots or even German Colonel Otto Skorzeny's glider and parachute commandos. All these leaders became world famous.

Bagnold shared the anonymity of the LRDG in wartime. He left the unit in the summer of 1941 to become Inspector of Desert Troops,(1) and shortly afterwards deputy signal-officer-in-chief, with the rank of brigadier. He was decorated for his achievement in forming the LRDG with the Order of the British Empire – an exceedingly modest award for his unique contribution to the security of the Middle East and the defeat of the Axis. As he left the LRDG in 1941, his name ceased to be associated with it thereafter, except by those who knew the whole story and the true story. Later writers tended to assume that the colorful LRDG had come into existence as though grown on a bush. Bagnold's personal indifference to publicity helped hide him to history and he was already half-forgotten when his LRDG brought off the classic climax to its career.

From his vantage point on the staff, Bagnold saw the LRDG trigger the end of the North African war, just as it had opened the Allied account in 1940. At Mareth in Tunisia, where Rommel made his final stand, a "left hook" was smashed home against the German forces that ended Axis hopes in Africa forever. This devastating knockout blow was delivered through country marked "impassable" on military maps. Leading the pulverizing stroke was Major General Sir Bernard "Tiny" Freyberg, who had given Bagnold the first troops for his patrols – back when Bagnold was known in Cairo for his "wild ideas." Freyberg had followed a route through "impassable" country found for him by a patrol of the LRDG.

After the war, Brigadier Ralph Bagnold retired from the army for good, the green tranquility of the Kentish countryside substituting for the golden wastes on which he found high adventure and fulfillment such as comes the way of few men. A busy and respected member of the British scientific community for decades, his fascination with the mysteries of natural physical processes was endless. He was a longtime consultant in the movement of sediments, beach formation and the like. In the words of Bill Kennedy Shaw: "Dry sand being difficult of access for him, he deals with wet mud."

Although unknown in the United States outside professional circles, in 1969 Bagnold became the first recipient of the G.K. Warren Prize, awarded by the National Academy of Sciences of the USA. The prize recognized his contributions to fluvial geology. In 1970 he was awarded the Penrose Medal of the Geological Society of America. He was further honored by the Geological Society of London with its Wallaston Medal in 1971, and the International Association of Sedimentologists recognized his achievements with its Sorby Medal in 1978. Bagnold died in London on May 28, 1990.

Bagnold's contribution to Allied victory remained but little known or understood even in his native England, which showered honors and historical affection on other desert and commando heroes. He could rightly be called the Allies' hidden hero of the North African conflict. Without his "mosquito columns," events in 1940-41 would have unfolded very differently. The mind boggles at the consequences of the seizure of the Suez Canal in the autumn of 1940 by Graziani's massive army. Rommel's army was only a third as large when, two years later, the Desert Fox nearly took Egypt and the canal.

With the slenderest resources, Bagnold and Wavell – each a visionary in his own way – aborted the disaster of Egypt and the Middle East being lost in 1940. In modern military history there has rarely, if ever, been a bluff of such magnitude. Certainly there has never been one pulled with such elegance and finesse as Bagnold's bluff, invested as it was with the conquering power of an idea whose time had come.

Note

1. Prewar companion and pilot Guy Prendergast relieved Bagnold as commanding officer (CO) of the LRDG, and happily flew half of the unit's "air force," which consisted of two WACO monoplanes. In the Bagnold tradition, Prendergast was the most mobile CO in the North African theater.

Author's Note

"Bagnold's Bluff" is a lightly edited version of a chapter of my book *Hidden Heroes*, published in 1971 by Arthur Barker, Ltd. (London). That was during the Vietnam War era, during which military men were widely disdained in America, and soldiers were sometimes even spat upon. The US market for military history was poor, and I went on to other things. Thus, Ralph Bagnold remained in the obscurity to which wartime security originally consigned him. The story of his remarkable strategic bluff in the fall of 1940 has therefore remained essentially unknown in the United States, and this essay is presented here to an American readership for the first time.

In August 1940, more than 200,000 Italian troops were massed on Libya's frontier with Egypt, poised to seize Cairo and the Suez Canal and thereby threatening the loss of the entire Middle East to the Axis. The strategic and geopolitical consequences of such a loss would have been incalculable. Nothing better points up the appalling weakness of the British defenders at that time than the fact that when Bagnold drew weapons for his patrols from the available supply, he found just three Vickers machine guns remaining as the total reserve for the entire Middle East. General Wavell used more guile than guns when he sent Bagnold's mosquito columns to raise hell on Graziani's supposedly secure right flank. History has certainly credited Wavell fairly for his February 1941 defeat of the Italians, but it was Bagnold's Bluff that caused wavering, apprehension and irresolution in the Italian command, despite the overwhelming numerical and material superiority of its forces.

The story of Ralph Bagnold strikingly points up how individuals can and do make a real difference in history. This extraordinary man has never been given proper public credit for his enterprising role in keeping Egypt and the vital Suez Canal from Axis hands, and thus in altering the war's entire course. During the war years, the story of Bagnold's dune-crossing route to inner Libya had to remain secret. And while his patrol force was successfully expanded, in 1941 Bagnold himself was promoted away to a lower-profile post, and thereby lost to history. While Wavell has received well-deserved acclaim for his victory over the Italians, Bagnold's key role in Wavell's strategic deception has remained veiled.

In writing this piece, I am grateful above all to Ralph Bagnold himself, with whom I had extensive correspondence. My long research on the Long Range Group convinced me that the most fascinating of the many stories the unit generated was that of its creation, from absolute zero, in beleaguered 1940. As a required Brigadier, Bagnold kindly assisted me with illuminating detail, which has appeared nowhere else, about this crucial start-up period. A gentleman of formidable intelligence, he kindly vetted my drafts of "Bagnold's Bluff," and provided valuable additions.

My good fortune, also as a young writer in the 1960s, was to become a correspondent of William Boyd Kennedy Shaw, former Major and LRDG Intelligence Officer. His book, *Long Range Desert Group*, published in 1945 by Collins (London), remains the basic work on the subject. An archeologist and Arabic scholar, Shaw had been one of Bagnold's dependable companions on the prewar expeditions into the Libyan Sand Sea, and beyond, which completed the primary exploration of that region. Through this ineffable gentleman, I was able to contact former Major Pat Clayton, another member of Bagnold's prewar exploration group who later helped him organize the LRDG, as well as retired Captain Richard Lawson, former LRDG medical officer, who kindly loaned me his precious photo negatives of the period.

I am indebted to all these late gentlemen for their generous and heart-warming aid.

Chapter 22

HISTORICAL ACKNOWLEDGMENTS

A Detailed Review of Participants in and Their Contributions to Etheric Rain Engineering since 1968

Ever since Trevor Constable (hereinafter TJC) began his involvement with etheric rain engineering technology around 1968, many people have aided and reinforced this development. Without their contributions, which have been as varied as the people themselves, the elegant airborne etheric rain engineering operations of today would have required many more years to realize. At key junctures, development may well have ceased were it not for interventions by interested, capable and kindly disposed people. Certain of these interventions, such as those of George K. C. Wuu, the Segreti Brothers, Irwin Trent and Dr. James O. Woods, decisively kept this project going when TJC had resolved to close it down, out of sheer weariness with the expense and strain of large, regional operations. No support of any kind is extended to work of this character by the great foundations with their tax-sheltered funds. They tend to sustain conventional, thoroughly safe ventures, rather than assist technical revolutions.

Cooperation keynoted all the assisting services that were freely given. No government research money was ever used or expected. Many of the individuals involved made substantial sacrifices of time, funds and skill. Scientists and other professionals contributed their knowledge and insight unstintingly, and without fees. Whatever the level of their involvement and conviction, everyone could see one overriding goal: human life would be elevated and advanced by the furtherance of etheric rain engineering. Most of those who helped also saw that beyond the break-in development of etheric rain engineering there loomed the wider, industrial developments of etheric technology, requiring no fuel and producing no pollution. This wider horizon fascinated and enthralled everyone involved. They all realized that we have to begin where we can, rather than dream idly of a perfect beginning. Those who shared our visions and have passed on are remembered with reverence and abiding gratitude.

These acknowledgments are recorded here so that through this book, persons who are newly aware of etheric rain engineering will understand the varied backgrounds of this technology's progenitors. One man's efforts could not possibly have sufficed. Motivations have been pure throughout. Cooperation by free human beings is the most powerful force there is for progress, crossing national, racial and political boundaries. Destined in the fullness of time to supplant the profit motive for advancing the human race, cooperation is needed even now on a world stage, to salvage this stricken planet from the ravages of unbridled avarice and disdain for Nature.

Grateful acknowledgment is extended to the following persons, for their contributions to progress. Not all of them were personally or directly involved in our particular activities but whether centrally or peripherally involved, all of them together helped produce the airborne etheric rain engineering of today. AEREO is their achievement. Evident and obvious is that these are human beings of high quality.

GEORGE K.C. WUU of Singapore is a visionary businessman, multi-millionaire and entrepreneur, who made possible the commercial availability of airborne etheric rain engineering operations today. Without his financial intervention in 1988, this activity would not have progressed beyond that time. The many historic operations conducted since then in Hawaii, California, Malaysia and the People's Republic of China would never have produced their harvest of know-how. These advances were purchased at considerable personal cost to Mr. Wuu. Born and raised in modest circumstances in Singapore, Mr. Wuu created the East Coast Recreation Centre there from virgin, freshly reclaimed land. He was given this opportunity by the Singapore government as a young ex-Singapore Navy officer, because the government of that dynamic meritocracy believed he had the drive and enterprise to succeed. They gave him his chance, and he justified their confidence. His rivals for the contract were some of the largest construction firms in Asia. His success with the Centre, which he has since passed to others, is now part of Singapore's vibrant modern history. George Wuu today is an active international businessman of wide interests and humanitarian bent. Airborne etheric rain engineering operations remain of special concern to him. Money from what he termed his "Peanut Fund" covered expenses for the primary pioneering of airborne etheric rain engineering in Hawaii: Operation Red Baron. As a "hands- on" business executive, Mr. Wuu has many times organized and executed rain engineering operations of his own, and has designed and fabricated equipment in Singapore. His intense practical interest in the technology included his own personal verifications of the basic airborne etheric rain engineering discoveries of TJC. The technology is commercially available today, because Mr. Wuu believes that it will be part of the world of tomorrow. He was most recently Chairman and Chief Executive Officer of Etheric Rain Engineering Pte.

Ltd., so for a time controlled the technology for which he was an energetic and indispensable midwife.

PETER LINDEMANN, D.Sc. is a great friend to etheric rain engineering, as well as possessing a remarkable level of personal sensitivity to etheric force. Peter is a brilliant multi-disciplinary engineer who has created many devices, ranging from therapeutic equipment to electrical generators, whose operational characteristics are outside of modern scientific theory – yet still function. His grasp of science and the history of scientific experiment are profound. Peter was a highly valued executive board member of Borderland Sciences for many years, and was a frequent contributor to the *Journal of Borderland Research.* Tom Brown has commented that he could always rely on Peter to be on the case with the latest insights on innovative scientific and technical matters that were under research around the world.

Peter and TJC met and became directly associated in 1990, although TJC had heard many praises sung of Peter by Webmeister Tom Brown, over the years. Seafaring almost invariably ensures that one is somewhere else when interesting people are around, and that is the way it was until 1990. At that time, Peter was living in Santa Barbara. He came to San Pedro – about a 100-mile drive – to collect some of TJC's obsolescent devices. Drought was severe in Santa Barbara and environs, and Peter felt that he might reverse that situation by "inheriting" equipment that TJC was no longer using and was actually about to dump. Peter hauled the assortment of tubes away, and in Santa Barbara, at the home of a friend, he organized this equipment into what we came to call "dynamic storage."

Peter is possessed of an astonishing sensitivity to etheric force, and he utilized this sensitivity to the ether to turn TJC's old tubes into a fully functioning force majeure in Santa Barbara. Between his collection of the tubes, and Peter's development of dynamic storage, TJC had failed dismally to interest the stricken city in a straightforward rain-engineering contract. With no money upfront, George Wuu would underwrite all costs, and we would, inside a year, transform the drought into a memory never to be repeated. The deal was to include five years of gratis consultations, and the installation of permanent facilities to deal with Santa Barbara's notorious "sundowner" winds, which have repeatedly pushed the city to the brink of incineration. Such an incident had occurred shortly before my 1990 presentation. To our intense disappointment, we were turned down, and the city chose to build a desalination plant, at more than ten times the cost. Santa Barbara is still "eating" the bonds involved, and since there have been no drought conditions since that time the plant has been decommissioned.

There is an excellent technical explanation for this astonishing turnaround. Peter Lindemann's "dynamic storage" entered the scenario at this time. He obtained a location for his etheric projectors at the house of a friend, located by good fortune, 5 miles due south of Gibraltar dam, whose reservoir was then "dust dry." Peter's superlative sensitivity allowed him to align the etheric projectors he was storing, so that the staple flow of etheric force from the west could not overrun the "stored" tubes.

Blocked locally by the stored equipment, the immense flow of ether from the west was shunted 90 degrees to the north, right over Gibraltar dam. There were soon consequences from this alignment. Two successive rainstorms delivered two massive rains to Gibraltar: both exceeding more than 11 inches while untargeted areas received about one inch. This sent Gibraltar Reservoir, over-night, from dust-dry to spilling over 100 million gallons of water an hour into Lake Cachuma, the main reservoir for southern Santa Barbara county. Gibraltar dam recorded the *highest rainfall in the region on both occasions*.

The geometry involved was exquisite, and complied with principles filmed in time lapse on the California high desert nearly twenty years previously by TJC, and the late biologist Robert McCullough. This is a real-world example of Peter Lindemann's superb sensitivity, and what may be achieved through that sensitivity as a purely practical matter. In the years since Santa Barbara and the Gibraltar dam, TJC has enjoyed immense benefit from his friendship and association with Dr. Lindemann. Among his many gracious gifts to rain engineering, Peter produced the fabulous P-Gun. This has made airborne etheric rain engineering 100 percent feasible, without chemical agents, or electric power in any form. What is now required is a political leader with the common gumption to utilize what the Cosmos has provided, via Peter Lindemann.

LOUIS A. MATTA served as Chief Engineer of SS "Maui" throughout the 1980s and until his early retirement from the sea in 1992. A professional engineer educated at the California Maritime Academy, Lou Matta's early personal interests were sacred geometry and the art of Tai Chi. The hard world of mechanical engineering and the operation of a hard-driven 33,000 horsepower commercial ship and all its complex sub-systems is more than enough for most men. For Lou, it was only a part of his life, because of the special character and breadth of his avocational interests. Tai Chi training allowed him to grasp immediately what etheric weather engineering was all about. He became vitally important to TJC's operational experiments aboard the Maui – truly a key personage in the entire unfoldment.

The two men conferred almost nightly while the ship was at sea, endlessly sketching, theorizing, proposing and then fashioning artifacts in the ship's engineering workshop. What was fabricated was immediately taken up to the flying bridge and tested in the dynamic crucible of the north Pacific Ocean – pristine and endless. Radar overviewed everything that happened. A more perfect setup for testing and experimenting would be hard to conceive. TJC duly plowed Mr. Matta into Dr. Rudolf Steiner's work, further enhancing his already invaluable assistance.

The sheer value of being able to take an idea aft to Lou Matta's office and discuss that idea with him is almost indescribable. This privilege was available for more than a decade as the "Maui" criss-crossed the eastern north Pacific. One of the burdens pioneers in new technology bear, is the devastating loneliness created by the soulless nature of our dehumanized, mechanistic culture. The steady presence of Lou Matta offset this. We both grew together into the new concepts and changed thinking that are coming to the world in the not-so-distant future – perhaps sooner than those stuck in the Old Knowledge realize. Lou Matta was many times during those formative years, a living witness to some stupendous weather events around the ship's vicinity that few others aboard recognized were engineered happenings. He has seen with his own eyes those towering, beetle-black barriers of moisture going up thousands of feet into the air ahead of the ship, and known exactly what he was looking at, and how it got there.

TJC cannot even begin to imagine how far back the development of etheric rain engineering would be, had it not been for the unforgettable aid, creativity and comradeship that came from Louis Matta. Warm-hearted, pure of motive, unselfish, kind, tolerant and moral, he is one of God's noblemen, sent to the times and places where he was needed.

The late DR. WILHELM REICH, M.D., who died in a Federal prison in 1957, was the major rain engineering pioneer and innovator with his inspired invention of the Cloudbuster. Protégé and sometime First Clinical Assistant to the eminent Dr. Sigmund Freud, Dr. Reich found the pathway from psychology and psychiatry to biology that eluded Freud and many other brilliant researchers. 20 years of professional scientific and clinical work preceded his discovery of a specific biological energy, which he termed orgone. A revolutionary break with classical formulations, this excited the professional enmity and irrational response that invariably awaits the true pioneer. His imprisonment, and the destruction by fire under court order of his books and experimental bulletins are an infamous episode in U.S. jurisprudence. Deliberately misrepresented, distorted, falsified and misunderstood to this day, Dr. Reich left behind not only his discoveries, path-breaking scientific books and experimental records, but more than 100,000 pages of unpublished manuscript – enough material for a hundred more scientific books. Perhaps a century hence, Wilhelm Reich will rank among the major heroes of the coming New Civilization. Few people now living are fully capable of comprehending the cross-disciplinary magnitude of his contributions to mankind's progress. Due to Dr. Reich's penetrating clinical findings on irrational behavior, psychopathic resistance to life-giving discoveries no longer need surprise anyone.

Modern scientific rain engineering, including everything achieved by the ERE group, is descended from Dr. Reich's fundamental discovery of the orgone energy, the name he gave to the ether of space.

Without the start he made in directly tapping the etheric continuum, there would have been nothing to build upon. While contemporary rain engineering equipment bears little resemblance to the classical Reich cloudbuster, the latter is contained in all that exists today – just as the Wright Flyer of 1903 is contained in the mighty Boeing 747. In the ERE group, we honor the memory of Wilhelm Reich, salute his achievements and benefit from his monumental legacy to the human race.

The late biologist, ROBERT MCCULLOUGH, was the link between Dr. Reich's work and that of TJC. Bob McCullough was educated at the Utah State Agricultural College where he earned a master's degree. For two years he was employed by Dr. Reich as a biological scientist. This work included personal assistance to Dr. Reich in pioneer weather engineering development and operation of the cloudbuster. These assignments included the famous expedition to Tucson, Arizona, half a century ago, that resulted in the summertime greening of the surrounding desert. McCullough contacted TJC when he saw infrared photographs in the latter's 1958 book on UFOs, that resembled objects McCullough had seen provoked by cloudbuster operations in the sky around Tucson. Close collaboration and friendship ensued, and lasted until McCullough's death in 1995. He was the technical adviser on the very first cloudbuster built by TJC and the late Dr. James O. Woods. TJC was fortunate and blessed to have the scientific guidance of Bob McCullough for three decades. In the course of this collaboration, information was exchanged primarily via dozens of audio cassette tapes. During most of this period, McCullough was a biologist with the Chemical and Biological Warfare section of the U.S. Army. Bob McCullough was a sensitive as well as a scientist. This made him not only a highly effective operator of cloudbuster equipment, but also a sound and solid scientific guide in the furtherance of this work.

EVA REICH M.D., physician and daughter of the great scientist, was a professional associate of Robert McCullough and her father, at Orgonon, in Maine. This outstanding professional lady guided the studies of TJC when the latter first investigated her father's work. Her careful supervision during this period ensured that Mr. Constable made a correct, non-tangential approach to the discoveries of

Wilhelm Reich, of which the cloudbuster is only a small part. In particular, Dr. Eva Reich ensured that Mr. Constable understood the neurotic rooting of social pathology, and its virulent consequences for any person pioneering life-force technology. This vital early guidance stabilized progress thereafter.

The late DR. JAMES O. WOODS was a highly valued research associate of TJC for more than a quarter century. The two men worked from 1957 until 1968 as a research team, in a novel project to objectify UFOs directly from the invisible state, using infrared film. NASA Shuttle mission STS75 in 1996, 39 years later, objectified numerous similar, invisible forms extensively on ultraviolet-sensitive videotape, including numerous examples of materialization into the ultraviolet and dematerialization out of the ultraviolet into other reaches of the invisible. The UFO research in 1957-68 acquainted TJC and Woods not only with the technical key to the UFO mystery, but also with the changed mental outlook essential to successful work with the ether. The infrared photographic project metamorphosed into the design, construction and operation of the first cloudbuster used by the two workers. Robert McCullough provided invaluable consulting guidance with all the cloudbusters built by the two men. The UFO research project provided extensive practical experience in working with the ether. Conversion of this experience to weather engineering was via a natural, functional pathway that has extended until today. Dr. Woods was a dear personal friend, a superb working partner in all phases of their joint efforts, and proved to be a talented, effective cloudbuster operator. A Doctor of Chiropractic, he was also for several years a clinical assistant to the late, redoubtable genius physician, Dr. Ruth B. Drown. Dr. Woods died prematurely in the 1980s, after a long illness, leaving a void that has never since been filled.

The late IRWIN TRENT was a veteran electronics technician already retired from the aerospace industry in California, when he met TJC in 1971. Mr. Trent's first acquaintance with Dr. Reich's work came as a patient of Dr. Albert Duvall, one of Reich's closest medical associates. TJC and Irwin Trent became dear friends as well as research associates for 30 years. "Irv" Trent provided unstinting, across-the-board support of the weather engineering work: financial, physical, theoretical, practical and emotional. He always kept a low profile, but was a steady and solid presence all through tough and demanding times. He was a bulwark in a time of personal tragedy. Irv Trent also personally published *The Cosmic Pulse of Life*, Trevor James Constable's underground classic book on energy matters closely related to rain engineering. *Cosmic Pulse* is now a recognized classic, many printings three decades later. Without Irv Trent, this book would have been suppressed by the publishing establishment that Trent personally despised. He also later helped subsidize publication of *Loom of the Future*, an interview book of novel format, with over 130 photos, recounting the history of rain engineering up to 1993. He was a longtime contributor to Borderland Sciences Research Foundation, and a valued member of that foundation. A successful investor who knew how to detect and shadow oligarchic machinations, Trent made a practice of quietly funding many people who suffered hardship because of their dedication to truth and justice. This man's many sterling qualities included his complete dependability. In the days of operations ashore in southern California, he could be sent to make equipment adjustments perhaps a hundred miles away. Although no longer a young man, Irv would always do exactly what he was instructed to do. He was therefore, and in many ways, a far-reaching additional pair of hands for TJC. Development of the "Skimmer," a simple yet effective piece of mobile rain engineering equipment, was his original design and inspiration. Compelling video exists of this device wringing rain out of massive high-pressure systems in the north Pacific. Irv Trent willingly drove gun-cars on the Los Angeles freeways, day or night – whatever was required. Father, uncle, brother, friend, sounding board, counselor and ceaseless experimenter, he was still experimenting a few weeks prior to his death at 86 in 2001. Irwin Trent played a large role in bringing etheric rain engineering into the world to stay. Humanity is in his debt, and that of his wife, Ethel.

The late WILLIAM GORDON ALLEN, Ph. D., was a graduate of Louisiana Tech who was long associated with Boeing in Seattle, Washington. Dr. Allen headed the interior electronics design group on the Boeing 747, during the birth of the revolutionary jumbo jet. His design group had responsibility for everything electronic from the flight deck aft. His major innovation on this project was development of the single coaxial cable for carrying all the electronic functions in the passenger cabin, such as programming and stewardess call-buttons, etc. This development made unnecessary the massive quantities of shielded wire previously necessary to fit out an airliner interior. Aside from the immense weight saving, the coax approach saved Boeing a billion dollars in costs on the first production run of the jumbo jet, and has been incorporated in airliner design since.

Dr. Allen also owned several radio broadcasting stations in the U.S. Pacific Northwest during his lifetime, and was a movie producer and author. He was long familiar with the ether, as it had been presented to the world by Dr. Rudolf Steiner and the European scientists who furthered Steiner's work. A deep student of Dr. Steiner, Dr. Allen was convinced that the science and technology of tomorrow would be based upon the physical reality and technological accessibility of the ether. He planned to obtain micron gold from the ether in the last project he pursued, post-retirement. Trevor James Constable was introduced to Dr. Allen by the late Dr. Ruth B. Drown, an extraordinary pioneer physician working medically with the ether, or life force. Dr. Allen was intensely interested in the development of etheric rain engineering, and provided cogent scientific and engineering guidance over more than a decade. He foresaw with clarity the inevitable development of airborne equipment and techniques. He judged this would be the most important development in the history of etheric rain engineering. His vision has been proved accurate. A great man, sadly missed.

The late FRANK DER YUEN was an internationally respected aeronautical engineer, airline executive and consultant, who was intensely interested in etheric weather engineering as a glimpse of tomorrow. He was a personal friend and business associate of TJC since the mid-1950s, and an ever-present source of wise counsel and engineering guidance. A graduate of Harvard and MIT, he was a leading engineer for Nationalist China's First Aircraft Factory during WWII, and later was vice president of Harlow Aircraft, a longtime consultant to Lockheed Air Terminal in California, Executive Director of the Honolulu Airlines Committee, and vice president of Aloha Airlines. He was one of the designers of the Honolulu International Airport, and founded the Aerospace Museum there. Mr. Der Yuen understood from personal experience that humanity can only tolerate small, incremental advances. Right after World War II, he designed the all-weather, self-propelled passenger loading bridges for airliners that are used in every major airport on earth today. When he took his proposal to Lockheed Air Terminal, they told him he was way ahead of his time. Ten years later, Lockheed presented him with his own drawings and asked him to help them build these bridges. Nowadays, every up-to-date airport in the world is equipped with them.

As a respected visionary with immense practical talent and ability, Frank Der Yuen was the catalyst in bringing innumerable processes and inventions to market, putting inventors and developers creatively together with financiers. Rarely did he receive any credit, acknowledgment or recompense for these vital services. Mr. Der Yuen was responsible for introducing George K.C.Wuu to etheric rain engineering, firing his enthusiasm, and bringing him together with TJC. Their long and fruitful association in developing this new scientific art to commercial viability is entirely due to Frank Der Yuen. In a poignant end to his noble life, Frank literally died in the arms of George Wuu as they boarded an airliner in Ontario, California – on one of Frank's famous loading bridges. This ineffable gentleman played a crucial role in bringing airborne etheric rain engineering operations to commercial feasibility. He was a true humanitarian.

COMMODORE KENNETH R. ORCUTT, USNR RET., was Commodore of the Matson Navigation Company's fleet, and master of their flagship, SS "Maui," throughout the major development of etheric rain engineering in the maritime mobile mode. His authorization of rain engineering equipment on the flying bridge of the Matson flagship was a considerable concession for a man in his position to make. This key privilege made possible a succession of advances in the art that would otherwise have required generations to achieve. The compelling technical saga facilitated by Commodore Orcutt's tolerance, is documented for posterity on numerous reels of time-lapse videotape. Some of this video is already in public release for the enlightenment of those who cling to the canard that there is no ether. A video dedicated to airborne etheric rain engineering operations, is available to responsible government officials on letterhead request to ERE Singapore.

A gentleman freely acknowledged by his contemporaries as blessed with a genius mentality, Commodore Orcutt was not a direct participant in our shipboard projects. He just quietly kept an eye on them. A trained observer of the first rank, and an aviator, nothing escaped him. His decades as a shipmaster and a student of meteorology, allowed him to recognize the colossal accretions of moisture we often raised over the ocean – with a high barometer – as something engineered and not natural. These observational powers extended to his profound understanding of radar as a navigational aid. These abilities allowed him readily to recognize the objective influence of the new kind of engineering that produced distinctive, geometric distribution of rains around the moving ship. Radar objectified unequivocally these geometric patterns. As a longtime amateur radio operator and an electronic technician, the Commodore also understood radar's technical constitution. A self-taught, expert computer programmer, Commodore Orcutt wrote the ship programs that now undergird what little remains of the sold out U.S. Merchant Marine. His computerized navigation programs essentially obsoleted tables and navigational methods in vogue for two centuries.

Commodore Orcutt was a former Panama Canal pilot, perhaps the most demanding of all nautical responsibilities. He once served as Hong Kong Port Captain for American President Lines, and was a yacht designer in Hong Kong. In his final assignment with the U.S. Naval Reserve, Commodore Orcutt taught ship-handling to prospective naval aircraft carrier handlers. An aircraft owner, skilled pilot and flight instructor, along with all his other stellar abilities and talents, he was the last person alive to be hoaxed by anything phony, but he remained open to something truly new. Through his insight and high native intelligence, Commodore Orcutt allowed something newborn to breathe. As a result, steady empirical work on shipboard proved decisively the physical existence and technical accessibility of the ether, thereby beginning an advance for mankind that no power on earth can stop.

Commodore Orcutt made yet another monumental contribution to etheric rain engineering: he endlessly suggested that we go airborne, even before it was technically feasible. He pushed the whole idea. As an aeroplane pilot and a shipmaster, he understood both the key media involved. He kept advocating that we turn our thinking toward airborne development. Technical advance to physically small geometric equipment with no chemicals or electric power in 1994, made airborne etheric rain engineering possible. All of Commodore Orcutt's expectations were dramatically fulfilled in actual operations. In the 21st century, the technology continues to follow the pathway first indicated by Commodore Orcutt. In his private life, he is a devoted, quiet humanitarian of many benefactions, most of them unknown even to his friends. This illustrious gentleman has no children, but thanks to him, Airborne Etheric Rain Engineering Operations have been born. These will bless the world's children long after we are all gone.

COLONEL WILLIAM A. SCHAUER USAF Ret. is the first pilot in the history of the world to fly Airborne Etheric Rain Engineering Operations (AEREO). He put in 31 years service with the USAF as a fighter pilot, and after retirement built a second career in general aviation in Hawaii. An elegant pilot and longtime president of the former Aviation Historical Society of Hawaii, "Willy" Schauer donated many professional hours to educating young Hawaiians on aviation. At the time Colonel Schauer agreed to fly the pioneer missions for airborne etheric rain engineering, TJC knew him only as having once sold an aeroplane to one of his friends: Commodore Orcutt of the SS "Maui." Having any pilot available who would tackle such a way-out venture, like engineering rain with hollow, empty tubes, was blessing enough. As the project turned out, Colonel Schauer proved to have truly phenomenal observational powers, which significantly aided the pioneering. Having spent 14 years with scores of deck officers aboard SS "Maui" – men who were professional weather observers – TJC was astonished to the point of awe by Willy Schauer's unrivaled observational talents. Colonel Schauer flew all the missions making up "Operation Red Baron," the designation of these pioneering flights, in 1994-96. After this, there was no doubt whatever that a major technical breakthrough had been made, and that rain engineering would henceforth progress primarily in the airborne mode. With a pilot of lesser observational talent and flying skill, AEREO would not be where it is today. When the world catches up to and grasps AEREO, the name of Colonel Willy Schauer will be among the "famous firsts" of aviation. Our supernal good fortune was to have him aboard: incomparably the man for the moment and the mission.

THE SEGRETI BROTHERS, GINO AND ANTHONY, helped support and sustain the rugged, pioneer years of etheric rain engineering from "Fort Zinderneuf" – Gino's home on the Mojave Desert 10 miles from Palm Springs, California. Gino's first contact with etheric rain engineering was as a curious quartermaster aboard SS "Maui," where rain engineering experiments were frequently carried out. While most other crewmembers ridiculed the rain project, Gino exhibited a gift for understanding something that might easily elude a Ph.D., and often has. With only one eye, and less than fifty percent vision in that one eye, he showed an incredible ability to recognize engineered rain formations. Without formal education, Gino's secret weapon was an uncanny, intuitive grasp of arcane processes – like the karmic awakening of ancient, dormant skills.

His interest in the entire project was intense and enduring. He wanted to continue with this work at his home on the desert –the toughest of all experimental environments – and to set up a permanent weather base there. He would be in cooperation with TJC's San Pedro operation on the California coast. He did this when a shipboard injury suddenly ended his lengthy U.S. merchant marine service, and he was "beached."

Gino Segreti (left) with Trevor Constable. Gino was Chief Weather Engineer at Fort Zinderneuf installation, 1985-2006 and shipmate with author (see pgs 169, 251). Equipment tuning for rain engineering is sometimes fine-tuned using a human sensitive – a person highly susceptible to detecting and receiving etheric and orgone energy. Gino was such a sensitive. Mathematical information is used for the basic tube proportions, however, the sensitive is able to detect the optimal length of the equipment either in person or or over long distances by phone. Also, senstives like Segreti may determine the proper direction and frequency of rotation of the Flying H and other rain engineering equipment for maximum performance on cloud busting and smog elimination. See also the following links on Odic Force:

https://en.wikipedia.org/wiki/Odic_force
http://web.archive.org/web/20131230041333/http://odicenergy.com/
https://en.wikipedia.org/wiki/Psychic

Many variants of geometric rain engineering equipment were tested and utilized at Fort Zinderneuf, faithfully supervised by Gino, who became an expert operator of this type of equipment. Joined later by his brother Anthony, also a U.S. merchant marine veteran, Gino held down "The Fort" for more than a decade. Anthony died in 1993. The loyalty, dependability and dedicated labor of the Segreti brothers are an example to all who merely wish for a better world. The thousands of hours they spent in the development of etheric rain engineering proved their willingness to work for a better tomorrow, all through the really tough years of this new scientific art.

COMMODORE C.C. WRIGHT, JR. was the senior master of the Matson Navigation Company, and in command of SS Maui, the Matson flagship, when TJC joined the ship on her maiden voyage in 1978. Intelligent, open-minded and a consummate gentleman, this former skipper of Matson passenger liners readily assented to trials of novel rain engineering equipment on the ship. He took a lively interest in results. Having spent thirty or more years studying maritime radar screens, he was immediately aware of the anomalous rain patterns induced by operation of rain engineering equipment under conditions of a high barometer and low humidity. He made helpful suggestions that contributed to verification of what he was seeing electronically, and permitted our first equipment installations on the ship's flying bridge. These were considerable concessions for a senior master to make, given the "off the wall" character of etheric rain engineering a quarter of a century ago. His interest was unflagging until his retirement in the early 1980s, when Commodore K. R. Orcutt replaced him. The scientific art of etheric rain engineering owes a debt of gratitude to Commodore Wright, still living in retirement as this is compiled in 2003.

The late GENERAL CURTIS E. LeMAY, USAF Ret., rendered unique personal assistance to Operation Clincher in southern California in 1990. Full details of this historic, seasonal victory over America's worst smog appear on the etheric rain engineering website. Their study is recommended for all who wish to base their judgments on results and facts. Operation Clincher was the most successful etheric engineering operation ever conducted by ERE personnel, but it would not have attained this status without the intervention of General LeMay. The Riverside, California region is a notorious accretion point for smog pollution, and that is where General LeMay lived his final years. Civic and commercial authorities there declined to assist Clincher, with backyard or rooftop bases. Upon learning that Operation Clincher needed operating sites in Riverside, General LeMay immediately arranged three, including one on his own patio. Subsequent to this decisive action, Riverside regional smog sharply declined from menacing levels, and the 1990 seasonal tallies there came in 11 percent under 1989. General LeMay's can-do action turned Operation Clincher into an across-the-board technical victory over smog, registering a 24 per cent seasonal reduction. General LeMay passed away prior to the conclusion of Clincher, but at the time of his death he was already arranging for etheric technology to be tested in clearing fog from the USAF's foggiest base. Men with the gumption and drive of Curtis LeMay are rare today.

The late HOWARD B. "SPUD" MORROW, one of the founders of the Morrow's Nut House national retail chain, was the first man ever to mention to TJC the notion that the weather could be controlled and rain engineered. TJC actually derided this idea when it was repeatedly brought up during desert camping trips in the early 1950s. Spud broached this idea often enough that TJC eventually ceased his objections and came to think that Spud might be right. As a millionaire businessman old enough to be TJC's father, Spud had kindly agreed to sponsor TJC when the latter came to America for permanent residence. In the mid-1950s, on camping trips, Spud introduced TJC to UFOs, then an intriguing and controversial latter day mystery. Out of this involvement came TJCs connection to the ether as the master technical key to the UFO mystery, as outlined in his *Cosmic Pulse of Life: The Revolutionary Biological Power Behind UFOs* (1976, updated 2008, Book Tree). From there, it was but a short step to using basic knowledge of etheric forces in accessing the ether directly and using it technically in rain engineering and kindred functions.

Nobody was more elated, delighted and consumingly interested in etheric rain engineering, until the end of his days, than Spud Morrow. He planted a seed back in the 1950s that came to bear abundant fruit. As a former aircraft manufacturer, veteran pilot and aviation enthusiast, Spud Morrow would have been ecstatic in 1994 to behold his vision going successfully airborne – his ultimate vindication. Etheric rain engineering operations over the years have produced some serendipitous wonders, but none more remarkable than those born of Spud Morrow's faith that engineered weather was not only feasible, but that it would come to pass. Spud Morrow is the godfather of AEREO.

CAPTAIN EDDO FEYEN, U.S. Merchant Marine, Retired, was for several years a TJC shipmate as Chief Officer of SS "Maui." This professional seafarer had been observing and reporting marine weather daily for decades. Eddo was an extremely acute observer, with unblocked perceptions. He was astonished at some of the monstrous rain accretions that were engineered into existence, virtually while he watched, despite a full array of rain-negating parameters. Interested, open-minded and intelligent, he soon grasped the geometric essentials of what was being done, and the simple principles involved. Before long, he could confidently predict how large, engineered rain accretions would move relative to the ship. These movements did not follow conventional patterns as predicted by the ship's collision avoidance system computer, but were obviously governed by a different and overriding geometry. On account of his intelligent interest, sailing with Eddo Feyen, was a productive experience for TJC, lasting until his own 1992 retirement. Promoted to Captain of another Matson ship in 1993, Eddo Feyen allowed TJC to install the immediate geometric forerunners of airborne equipment on this other ship for crucial and successful provings. Captain Feyen's reports, which TJC knew he could trust, saved us at least a developmental year prior to Operation Red Baron in Hawaii.

BEN GREGORY, who will be referred to herein simply as Greg, was an ex-U.S. Army soldier on disability, who resided across the street from Fort Zinderneufin Desert Hot Springs, California – our longtime desert base. Intrigued by the bizarre goings-on at the Fort, Greg made our acquaintance, and rapidly developed a consuming interest in weather engineering. This was by no means a one-sided happening. Greg brought with him a stunning, innate aptitude for working with etheric rain engineering equipment, which our group was quick to recognize and develop. Greg's aptitude was actually the consequence of a near-mortal encounter, while in the Army, with spinal meningitis. His medical treatment had included, incredibly, more than THIRTY spinal taps, the scars of which lined the flesh along his spine. Etheric weather engineering relies absolutely on the existence and geometric accessibility of etheric force in the atmosphere. This same etheric force is what animates our physical bodies, which otherwise descend into mineral inertia – as in death. A long-known principle of esoteric science and medicine is that the ETHERIC BODY each of us possesses, and which makes us living beings instead of corpses, is forced out of occlusion with the physical structure by chemical medicines and drugs. In Greg's case, about three-dozen spinal taps had resulted in a permanent LOOSENESS of his etheric "double body." This looseness of his etheric double resulted in Greg's extreme sensitivity to movements of the ocean of etheric energy in which we all live. Etheric weather engineering is the art and science of geometrically manipulating that etheric energy ocean and its many currents, tides and natural characteristics. Greg had been surgically made (unwittingly by Army surgeons) into a translator for such forces.

When Greg came to us, he knew nothing of such things. He was skilled as a motor mechanic, electrician, carpenter, plumber, welder and tiler and not given in any way to mysticism. He nevertheless exhibited from his first contact with weather engineering equipment, a truly phenomenal ability correctly to align and orientate such equipment. All this developed with rapid and impressive results.

As he was familiarized with basic knowledge of etheric force and its key role in weather formation, Greg became increasingly effective as an equipment operator. When he put a piece of equipment into a resonant alignment or orientation, his "loose" etheric double would react with a sensation akin to electric shock. The equipment was then "on tune," and objective physical results would ensue in local, visible weather phenomena. This invaluable sensitivity was understood by TJC, and carefully cultivated, so that in time: Greg became the most effective operator we had. TJC would often contact Greg by phone, from a hundred miles away, and have him through this link, tune TJC's equipment. There is no "here" or "there" in the etheric world, only tuning! When P-guns were developed, Greg had the sensitivity necessary to put these arcane structures on tune. The power of the P-guns is such that, even in their mini-versions, finding resonance would often send Greg flat on his back or evoke a loud cry of pain!

As Dr. Wilhelm Reich so aptly expressed it: "One must live things to know them."

The extreme significance of sensitivity was long recognized by TJC, through his studies of the work of Baron Karl von Reichenbach. This Austrian nobleman, one of the founders of the modem German chemical industry, had used his immense wealth to finance detailed investigations of etheric force, termed "od" in his writings and scientific papers. The Baron thoroughly understood the importance of "sensitives" to advancing human knowledge of etheric force. Modem America exhibits little interest in such advances, being currently seduced by the misuse of science and technology to destructive ends.

People like Ben Gregory should be cherished and sequestered as valuable human instruments for charting our course into the earth's tomorrows. What ERE Singapore of today offers to a droughted world, owes a great deal to Ben Gregory, and to the unique medical and karmic circumstances that led him to us. ERE and especially TJC, thank Greg for many years of help and cooperation toward a better world.

Chicago-born THOMAS J. BROWN has been involved with the development of etheric rain engineering for decades, and is today the webmaster for this ERE Singapore site. Now a businessman in New Zealand, Tom Brown has labored long and mightily in that most difficult of all assignments: presentation of new ideas and conceptions to a world determined to go to hell. Formerly Director of Borderland Sciences Research Foundation in California, during its most active research and publishing period (1985-1995) he edited while there a new kind of book in the form of a gigantic interview. *Loom of the Future* is a heavily illustrated book, with 130 photos keyed to the text. *Loom* dynamically presents the story of etheric rain engineering development under TJC up to 1993, and is a masterpiece of constructive editing by Tom Brown and his late wife Alison. There is probably no one alive who better understands the numerous, complex travails involved in midwifing etheric rain engineering than Tom. His grasp of a vast spectrum of New Knowledge reinforces his solid editing and writing talents. TJC is personally indebted to Tom Brown for years of dedicated support and friendship, including his deft handling of nitwit critics. His design of the rainengineering.com website is but the latest of his many generous deeds in the service of others.

APPENDIX

General Jimmy Doolittle greets TJC at an air power symposium in Long Beach, CA. General Doolittle wrote the Introduction to "Fighter General", the Constable and Toliver biography of Germany's General Adolf Galland,Doolittle's main opponent in WWII

Colonel Raymond F. Toliver USAF Retired, appears here on General Doolittle's left hand. Toliver co-authored four aviation books with TJC.

J. H. DOOLITTLE
P.O. BOX 566
PEBBLE BEACH, CA 93953

September 21, 1990

Trevor J. Constable
P.O. Box 746
Hauula, Hawaii 96717

Dear Trevor,

Thank you for your letter of September 17 and for the informative tape on primary energy weather engineering. Reviewed the tape with John and found your efforts most interesting. Imagine producing rain from blue skies within 2 hours. Every good wish for further progress in the ether project.

All is well here. Enjoyed a summer of visits from grandchildren and great grandchildren.

All the best to you and yours for much happiness and good health.

Very sincerely,

Jim Doolittle

J.H. Doolittle

One of America's legendary aviation heroes, General Jimmy Doolittle left a message for all mankind in one of his last TV interviews:

"LEAVE THE WORLD BETTER THAN YOU FOUND IT."

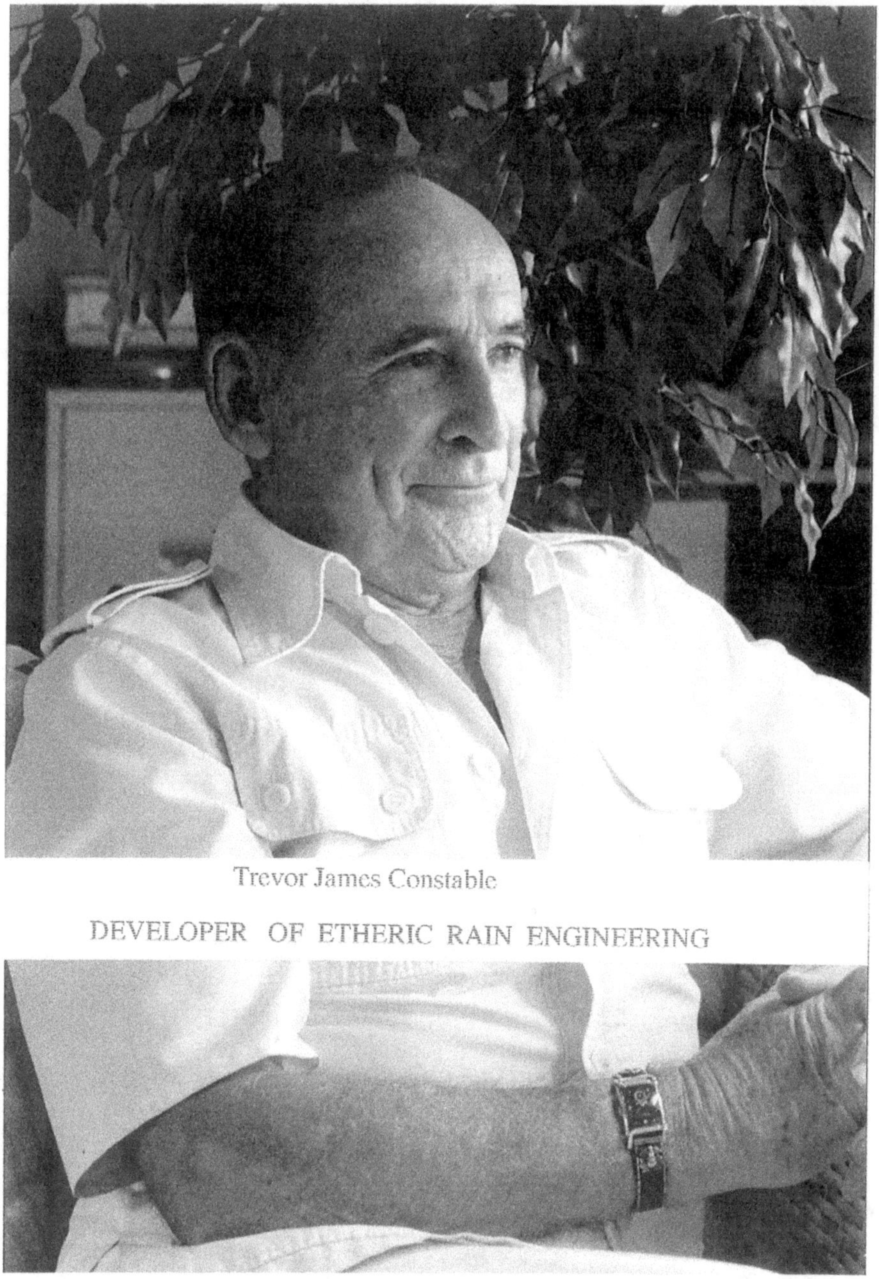

Trevor James Constable

DEVELOPER OF ETHERIC RAIN ENGINEERING